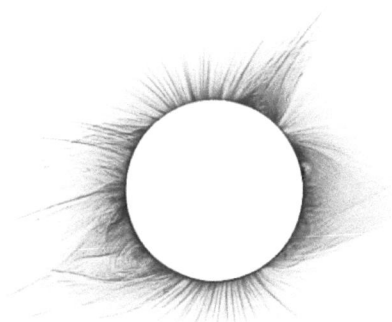

An **ECLIPSE** to REMEMBER

Steve and Christine Rosenow

Copyright © 2017 Steve Rosenow

All rights reserved.

ISBN: 1979889740
ISBN-13: 978-1979889742

DEDICATION

This book is dedicated to my wife, Christine. There is nobody else on this pale blue dot, I would rather share this experience this with.

CONTENTS

	Acknowledgments	V
	Foreword	VII
1	AN ECLIPSE TO PLAN FOR	8
2	STARTING TO GET SERIOUS	22
3	THE EXCITEMENT BUILDS	30
4	ARRIVING IN MADRAS	44
5	"WHO'S THAT GOOGLE GUY?"	64
6	TOTALLY ECLIPSED	75
7	FAREWELL MADRAS,	92
8	AFTER THE ECLIPSE	101
	GLOSSARY	106
	SOLAR ECLIPSE TERMINOLOGY	108
	ABOUT THE AUTHORS	110

ACKNOWLEDGMENTS

I feel that this book would be incomplete and an utter pile of rubbish, without the generous contributions of a few people with whom I feel a great sense of gratitude. First, I would like to acknowledge the following individuals:

Fred Espenak. Without his expertise and years of eclipse experience, this book would not have been possible. His *Mr. Eclipse* and *EclipseWise* websites inspired our trip, and ultimately laid down the groundwork for what started my "eclipse fever."

Xavier Jubier. Without his interactive maps and interactive Baily's Beads charts (using the Watts-Kaguya prediction method), I wouldn't have known where the best spot would be to see that phenomena.

Michael Zeiler. His unique Great American Eclipse memorabilia became our "uniform" while down in Madras, Oregon. I still regret not getting every piece of merchandise I could. His maps became quite a favorite with some of us there on that ridge.

Secondly, I'd like to acknowledge my coworkers. Trey, Leah, Paula, Bear, David, Hannah, Tracey, Jayden, Triston, and Jake (plus everyone else I can't mention because the list would literally be a mile long!). Thank you for putting up with more than eight months of nothing but eclipse talk, and for me continuing to talk about it! It's hard not to get excited about what literally is a once-in-a-lifetime event for some. And for me, as an astronomer, I felt this event was more than that. For many astronomers, an eclipse is a Holy Grail experience. I owe each of you a great sense of gratitude. Even though all of you could've told me off, but instead didn't, means the world to me. Thank you so much for that.

I would also like to thank my family, especially my mother. Putting up with me talking about the eclipse for the better part of the last decade means a lot to me. I love you, Mom.

I'd additionally like to thank my close friend Meg McDonald. Without her, one amazing eclipse video wouldn't exist. Her relentless pursuit of perfection in the process of that collaboration, resulted in one of the best joint projects I have ever been a part of.

Scott Sistek at KOMO TV in Seattle, Washington, deserves a huge "Thank you" for featuring my eclipse image in the moments after totality. Being the first person in the KOMO "Legion of Zoom" to take a photo of the eclipse, and have it featured on the KOMO website literally as it happened, was a true honor and privilege.

Lastly and most importantly, I would like to thank my beautiful wife, Christine. Without you, my love, this entire experience wouldn't have been possible. Planning and executing this eclipse trip over the course of the last year put a lot of strain on both of us. Now that we can look beyond that, it turned out to be one hell of a memorable trip. Not only that, it made for an extremely awesome and belated honeymoon!

I wouldn't have missed this for the world, and seeing you experience true joy at witnessing this eclipse meant so much to me.

I love you in more ways than one.

AN ECLIPSE TO REMEMBER

FOREWORD

The idea of writing a book about the eclipse – much less our experience about it - wasn't something I would say I came up with on my own. I'd even go as far as saying that initially, I had no plans of ever writing one.

In my day job, I work as a photo lab technician and photo equipment sales associate for a major retailer. By night, I am a semi-professional photographer, with some esteemed accomplishments to my name.

In my day-to-day routine in the photo lab, I come across a lot of unique and interesting people, many of them regular customers, coworkers, and vendors who grew impressed with my tenacity and my desire to witness the 2017 total eclipse.

It had been a few weeks after the eclipse, when I was telling the story of our experience to a couple vendors and a customer. All three said in almost deadpanned unison: "Hey, you should write a book about it!" Surprisingly, one of my coworkers agreed with them.

At first, I thought it was a joke, and implied they must be joking because, hey, who's ever done THAT?

When I asked if they were joking, they said they were serious.

Then, a sudden epiphany hit me: It's not often that in the pursuit of witnessing a total solar eclipse, you end up getting chased away by overzealous ranchers and a 2,000-pound wild bull, all the while you're trying to maintain your sanity amid rising anticipation of seeing a once-in-a-lifetime celestial event. When you're watching a total eclipse for the first time, your senses can easily become overloaded. Witnessing the landscape and sky undergo drastic changes within a short period of time, is an experience one will never forget. The same goes for seeing the solar corona, its fine wispy details shining like moonlight in daytime skies.

Even crazier, was the notion of playing Rock-Paper-Scissors with the car next to you, because you're stuck in endless gridlock on the way home for fourteen grueling hours, or witnessing the same stretch of road become an impromptu campground.

This book tells that story

Steve Rosenow
Shelton, Washington

AN ECLIPSE TO PLAN FOR

PREFACE ON SOLAR ECLIPSES

Eclipses, especially those of a total solar variety, have been fascinating mankind for time immemorial. To early mankind, eclipses were seen as a sign of impending doom. Many thought the sun's light was being extinguished. Others would view them as a bad omen, or a sign of impending change. Ancient Vietnamese people believed that a solar eclipse was caused by a giant frog devouring the sun, while Norse cultures blamed wolves for eating it.

In ancient China, a celestial dragon was thought to lunch on the Sun, causing a solar eclipse. In fact, the Chinese word of an eclipse, chih or shih, literally means to eat.

According to ancient Hindu mythology, the deity Rahu is beheaded by the gods for capturing and drinking Amrita, the gods' nectar. Rahu's head flies off into the sky, and then swallows the Sun causing an eclipse.

AN ECLIPSE TO REMEMBER

Korean folklore offers another ancient explanation for solar eclipses. It suggests that solar eclipses happen because mythical dogs were attempting to steal the sun. Traditionally, people in many cultures get together to bang pots and pans and make loud noises during a solar eclipse. It is thought that making a noise scares the demon causing the eclipse to run away. Some ancient civilizations would cast flaming arrows at a totally eclipsed sun, in vain attempts to reignite it.

Modern superstitions exist, as well. In many modern-day cultures, it is believed that total solar eclipses harm pregnant women and their unborn children, and in those cultures, it is asked of women to stay indoors during one. In India, some participate in fasting rituals, due to the belief that food consumed during a total solar eclipse is seen as impure and poisoned.

Mankind's understanding of solar eclipses and the dynamics in which they occur, have only began to be understood within the last few centuries. Advancements in eclipse science, including the accurate forecasting of where eclipses occur and when, have been happening only within the last century or so.

In 1887, Austrian astronomer Theodor von Oppolzer compiled the first major prediction catalog of solar eclipses. Titled *Canon der Finsternisse*, it was a set of 160 eclipse tables and maps spanning over three millennia from 1207 BC to 2161 AD. *Canon der Finsternisse* was a major breakthrough in eclipse prediction, and the prediction for 2017's eclipse was featured both in a map of eclipse paths, as well as a table of contact times and locations. Despite the significant breakthrough in eclipse predictions, *Canon der Finsternisse* was still flawed as only the start, middle and end of the eclipse were accurately illustrated; only curves were drawn through those to predict the path of totality. In many cases, such as 2017's prediction, the path of totality was several tens of miles off in terms of accuracy.

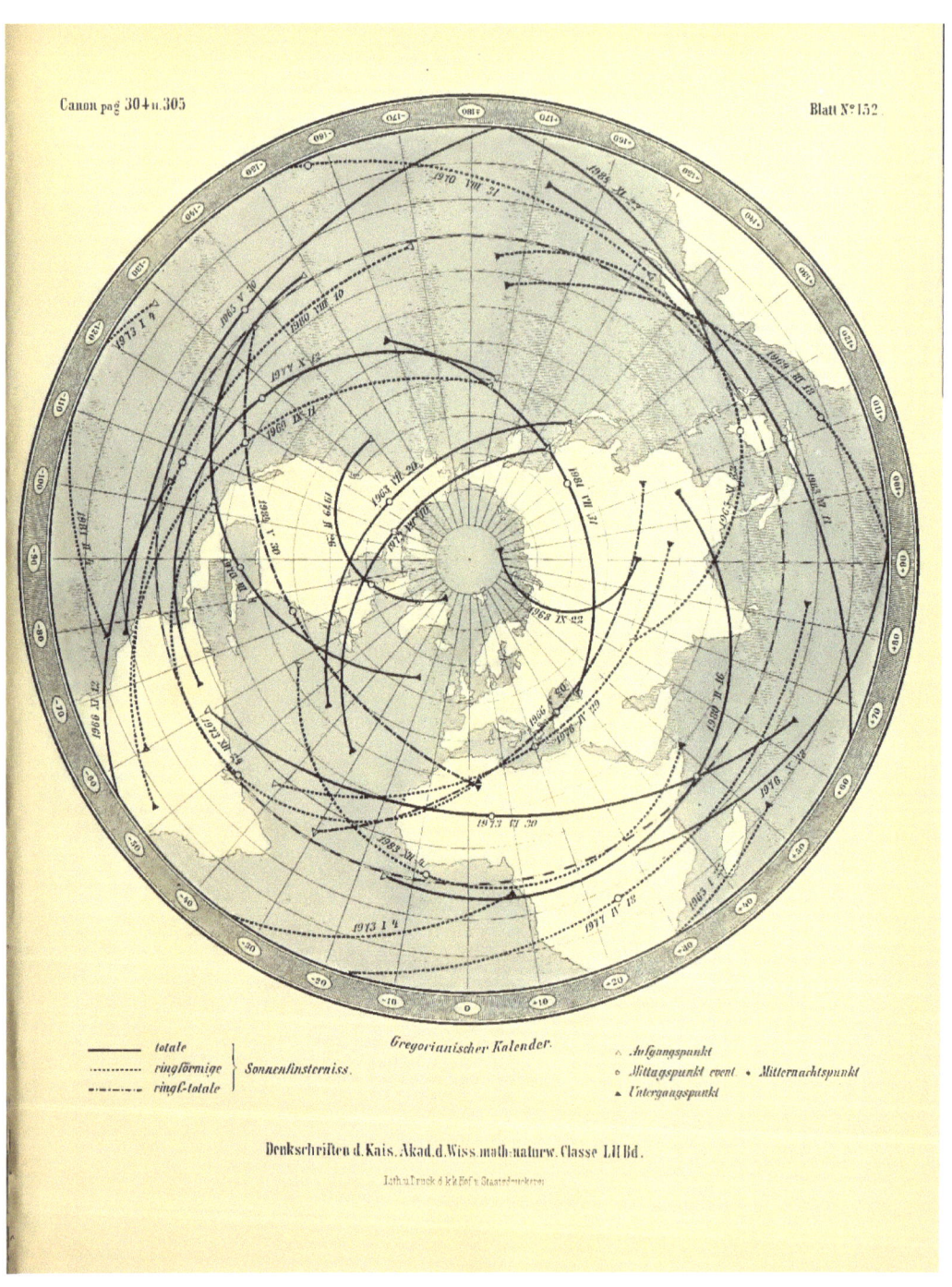

AN ECLIPSE TO REMEMBER

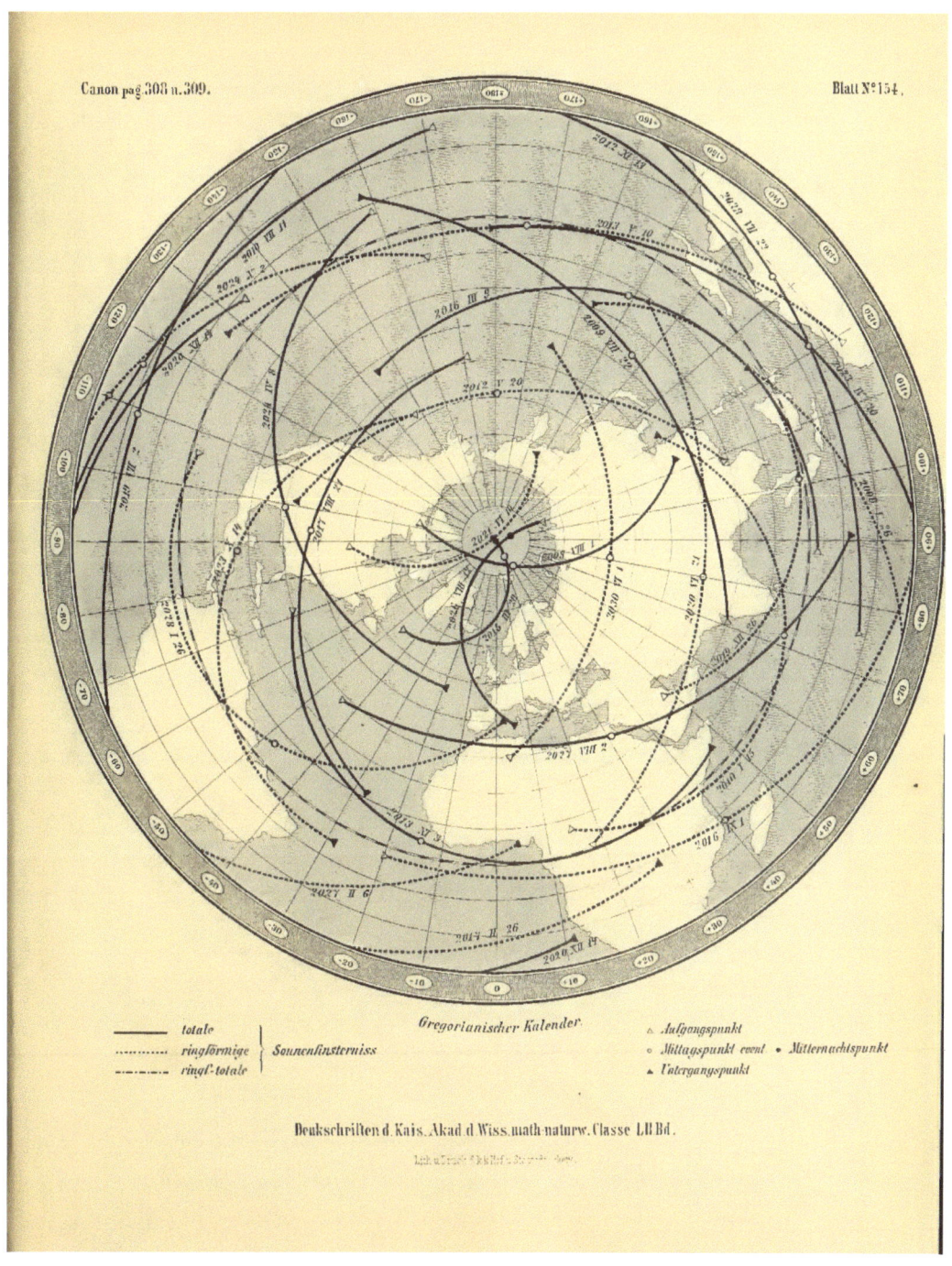

TOP AND PRECEDING PAGE: Selected prediction calculations from Oppolzer's *Canon der Finsternisse*, highlighting the predicted paths of totality for the 1979, 2017 and 2024 eclipses.

In 1932, scientists began working with improved accuracy on their prediction models, and in August of that year the New York Times ran a brief mention of the 2017 eclipse. One astronomer even hinted that it might be the "only good chance for the next 85 years," citing weather probabilities for the March 1970 eclipse.

It was not until 1966 however, when *Canon of Solar Eclipses* was published by Belgian astronomers Jean Meeus, Carl Grosjean, and Willy Vanderleen, that the accuracy of solar eclipse prediction, as well as the accuracy in predicting the paths of solar eclipses, was refined.

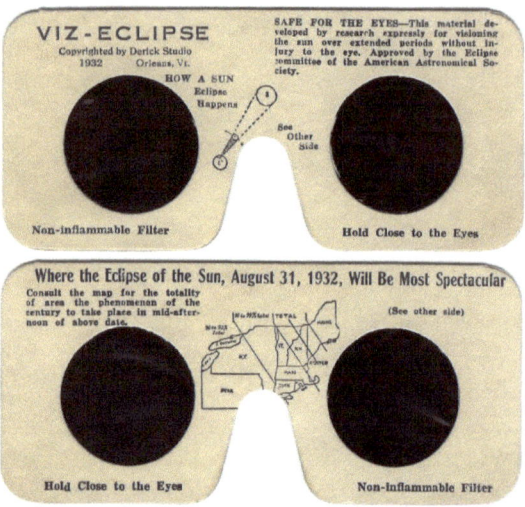

Early examples of eclipse viewing glasses. These were made for the August 31, 1932 eclipse, which dazzled eclipse watchers throughout the northeast United States and Canada.

Images courtesy of the Plaistow Historical Society, Plaistow, New Hampshire.

In *Canon of Solar Eclipses*, the path of totality for the August 21, 2017 eclipse was remarkably accurate, especially since it was the first major eclipse catalog to be compiled with a computer (an IBM 1620 Data Processing System). As a result, many points were calculated for each eclipse path, allowing for a far greater degree of accuracy. In the mid-1980s, NASA astrophysicist Fred Espenak, would further refine the science of eclipse prediction with his *50 Year Canon of Solar Eclipses*.

The accuracy of the path of totality and duration of totality at various points along the eclipse path for the 2017 eclipse, ironically were refined during the eclipse of February 26, 1979. With the exception of some areas in eastern Washington and northeastern Oregon, much of the Pacific Northwest was blocked from seeing the eclipse by way of clouds from an advancing storm front. Taking advantage of eastern Oregon's typically-clearer weather, scientists from the University of Oregon staged an expedition to the small city of Madras to study the eclipse. As a result of that team's expedition, the 1979 eclipse was groundbreaking in its discovery.

AN ECLIPSE TO REMEMBER

Among their findings, astronomers discovered that the terrain variations in the lunar limb affects how long totality lasts for a certain area, in addition to discovering that Earth's own terrain has an effect on where the path of totality falls. For example, their findings concluded that for an observer in the northern hemisphere, for each one thousand feet in elevation, the path of totality at any given spot on Earth shifts approximately two thousand feet or more to the south.

For the 20th century, 1979 was the last year in which the North American continent experienced a total eclipse. As ABC News covered the event, the opening narration by announcer Bill Rice declared, "A total eclipse of the sun, visible today over North America, for the last time in this century." Coverage of the event included cameras in Goldendale, Washington, Portland, Oregon, and in Helena, Montana. In Portland, the view of the eclipse was clouded out, with skies looking like midnight at just after 8:13 a.m. More than a hundred miles inland at the Goldendale Observatory, views were spotty of the eclipse. Television images showed clouds obscuring the views right as totality commenced. As the eclipse exited the United States in Montana, skies were clear enough in the city of Helena, that ABC was able to show the entire duration of totality. Soon after, the eclipse would exit the United States, and then travel into central Canada.

Thirty-eight years would pass before North America would be graced with another eclipse. In fact, as ABC was closing out their half-hour special report on the eclipse that morning, anchor Frank Reynolds uttered a promise to viewers: *"And so that's it, the last solar eclipse to be seen on this content, in this century. And as I said, not until August 21, 2017 will another eclipse be visible from North America. That's 38 years from now, and may the shadow of the moon fall upon a world at peace. Of course, ABC News will bring you a complete report on that next eclipse..."*

Ironically, thirty-eight years was also the amount of time that passed between the August 31, 1932 eclipse and the March 7, 1970 total solar eclipse (In fact, the 1970 eclipse was the same event that Carly Simon wrote about, in her hit song *"You're So Vain"*).

DEVELOPING AN INTEREST

The July 11, 1991 eclipse as seen in a simulated view from Shelton, Washington. Simulation provided by Stellarium (GNU/Public License).

My interest in solar eclipses began as an eleven-year-old in 1991. On July 11th of that year, a total eclipse of the sun occurred. The path of totality in that eclipse began some 1,300 miles west of Hawaii, intersecting the Big Island before moving east. As it approached the peninsula of Baja California, it turned south, passing over Cabo San Lucas.

Afterwards, it then continued onwards into portions of southern Mexico, Central and then South America. In my home state of Washington, it was experienced as a partial eclipse with less than 21 percent of the sun's disc blotted out by the moon. Being quite the crafty one, I used anything I could to see that eclipse. The closest objects nearby that would work, were a stack of

5.25" computer discs from an old Commodore 64. I reached for a couple, and as it turned out, they made a pretty decent naked eye solar filter.

The eclipse was one of the top stories on the local news that night, and it was mentioned that the Pacific Northwest was going to experience an eclipse of its own twenty-six years later on August 21, 2017. To an eleven-year-old in 1991, waiting that long to see another eclipse seemed like an eternity that would never end. I just couldn't wait for 2017 to happen.

TOP: My first telescope. Photo taken in October 2007. NEXT PAGE: The Lunar Eclipse of August 28, 2007

BECOMING AN ASTRONOMER

As I entered adulthood, I picked up the hobby of amateur astronomy after witnessing a total lunar eclipse on August 28, 2007. That night, I set up my first telescope, which was a small and portable Meade ETX-80. I purchased it at a camera store in Olympia, Washington a couple weeks prior. It was the first telescope I ever bought brand-new, and it came with an electronic touch-pad control. With more than 30,000 celestial objects in its automatic database, I used it every chance I got.

Another advantage to it, is that it was considered a 'Backpack Observatory' by Meade. It came with a small backpack and a relatively lightweight tripod. At the time, my cameras were a Konica 35MM SLR and a Minolta 'bridge' camera. That night, with the Minolta digital camera mounted to a tripod, I took my first eclipse photo. That first shot, a noisy, grainy capture of that lunar eclipse had me hooked. It was at that point, that I had made it my goal to witness the 2017 eclipse.

As the years passed, my astronomy hobby grew more serious. Part of my interest in astronomy was a long-held interest in space and space exploration. Another part of it, was my involvement in photography. A year or so after the lunar eclipse, I picked up photography (which later turned out to be a profitable-at-times business), and through both I learned that there was a lot of steps involved in telescope-based deep sky photography. I also picked up a greater understanding of when – and how – eclipses occur.

Incidentally, part of the reason I picked up the hobby of photography was due to that lunar eclipse, and because later that year, a short-period comet by the official designation of 17P/Holmes would literally explode in the night sky. In fact, Comet 17P/Holmes was the first time I'd ever do astrophotography with the aid of a telescope. As night fell, I loaded a 35MM SLR with 400 ASA Kodachrome Film, then zip-tied the camera with a 135mm fixed-focal length lens onto the back of the ETX-80 with the intent of trying to capture my first comet. A roll of film was shot that night and out of 27 exposures, the Comet Holmes image was the one that stood out the most.

AN ECLIPSE TO REMEMBER

Comet 17P/Holmes in October, 2007. The bright star immediately below it, is Mirphak (or Alpha Persei) in the constellation Perseus. Photo is a digitized scan of a 35MM negative.

It wasn't too long, before my telescopes grew larger in aperture as well. Five years later in 2012, my growth in the hobby of astronomy exploded. In June, I bought a larger Meade ETX telescope: the Meade ETX-90. It was a different beast compared to the Meade ETX-80 refractor I had previously used.

Unlike the ETX-80, which was an achromatic refractor on a computerized mount, the ETX-90 was a compound – or 'catadioptric' telescope housed in a roughly-identical fork-style mount. Its primary mirror was 4.5" in diameter, and it had a small, silvered secondary mirror on its meniscus 'corrector' lens. The ETX-90 was considered a Maxsutov-Cassegrain telescope (named after the inventor of the optical design) and it came with the sharpest planetary views I had ever witnessed through a telescope.

However, attempts at deep sky photography failed with the ETX-90. Those failures were mostly due to the fact that it was incapable of carrying the weight of the new Nikon DSLR I'd just bought a few months prior, and also because Maxsutov-Cassegrains carry the disadvantage of being optically slow performers for deep sky work.

Two weeks after I bought the ETX-90, a once-in-a-lifetime celestial event took place that I was fortunate to grab a few captures of. From our vantage point on Earth, Venus transits the sun in pairs on an eight-year interval, followed by more than a century between pairs. In each eight-year-pair, it is not unusual for the date of the latter transit to occur within a day of the anniversary of the former. The day the transit occurred was June 5th. The weather on that day, was less than ideal. For most of it, clouds stole the show. After capturing a half dozen or so photos of the transit, it had me more than eager for the 2017 eclipse.

One of several captures taken during the Transit of Venus on June 5, 2012. Venus is top center. The other 'specks' in the image are a complex grouping of sunspots, with each sunspot being the approximate size of Earth.

A month after the Transit of Venus, I decided it was time to upgrade from the ETX-90 to something much larger. I had found a 14-inch Meade LX200 Classic on Craigslist, but there was a small catch: I would have to drive to Springfield, Oregon to grab it. At 11:30 a.m., I set out on a five-hour drive from Shelton to Springfield. I was excited. However, the excitement quickly turned to disappointment when I discovered that the seller, who had promised to hold it for me, had sold it less than five minutes before I arrived. Not to be deterred from my goal of buying a Meade Schmidt-Cassegrain, the search continued.

AN ECLIPSE TO REMEMBER

After expressing my frustration to the seller, I left and stopped off at a rest area north of Eugene. While there, I spent an hour perusing the local Craigslist. It was about six o'clock at night when I found a potential candidate. After calling the number on the listing, I found out that the telescope would cost me half the cost of what the LX200 would've been, and it came with a lot more in the way of accessories. The next day, I arranged to meet the seller of that telescope at the marina in Allyn, Washington, a drive not far from home.

My Meade LX5 is seen in this September, 2012 photo. The keypad at the base of the telescope was a computer-assisted aiming device. Once polar alignment was achieved, signals from two pulley-driven encoders (one on each axis) would guide the user to a pre-selected deep sky or solar system object.

This was the innovative precursor to modern "Go-To" telescope technology. The hand controller near the declination lock knob and setting circle assembly, was for fine-tuned manual guiding when used for deep sky photography. It also served as a variable-rate tracking control.

After a brief demonstration of the telescope by the seller, I bought my first Schmidt-Cassegrain telescope. It was a classic, 1980s-era Meade LX3/2080 model, and it came from a seller in Gig Harbor. It was a monster compared to my ETX-90, with an 8-inch aperture mirror and an optical tube nearly as large as an antique milk can. With the LX3 being from the pre-digital control era, I lost the automatic celestial database. I figured it was a sacrifice that was well worth it compared to the sheer size and gain in light-gathering power. It also meant my images of the sun – and the 2017 eclipse – would be so much crisper. The accessories it came with, would also make an astronomer smile with glee. Included with the telescope itself, were over a half dozen eyepieces. Additionally, it had every filter imaginable for deep sky imaging and observing, a lunar filter, plus a dozen camera accessories. I would spend the next few years learning the night sky, and learning patience.

Three years before the eclipse, I guess you could say I got lucky. I was in a position to upgrade from my 8-inch Meade to something larger, and I was feeling what astronomers call "aperture fever." Aperture fever is a constant desire to own a telescope larger than the one currently owned. It often happens when you experience looking through a larger telescope for the first time after looking at the same object through a smaller one. Larger telescopes carry much more light-gathering power (which means that the details of much fainter objects become more readily visible), and their images tend to be a lot sharper due to increased resolution. A lot more detail is capable of being seen through a larger telescope, as well. For instance, the ring detail of Saturn or the cloud band detail on Jupiter is more readily visible in a larger telescope than a smaller one.

In late 2013 the aperture fever hit hard, and I picked up a 10-inch Meade LX6. When compared to my 8-inch SCT, it looked like a giant. The digital 'Push-to' electronics on it worked half of the time, but that was an acceptable trade-off when you consider that a ten-inch telescope has over fifty times the clarity of an eight-inch model.

Since the electronics of the LX6 were at times non-functional, I decided it was time for another upgrade. In February, 2014, I purchased a decommissioned ten-inch LX200 fork mount assembly. Once it arrived, I swapped my LX6/2120 optical tube assembly onto the newer, and much larger unit. The new LX200 mount came with an electronic hand controller and a built-in computer housing more than 30,000 celestial objects. It was also extremely heavy and at first, difficult to operate due to its sheer size and weight. Like the ETX-80 and ETX-90, it featured "Go-To" technology, where I could select a night sky or daytime object (such as Venus or Jupiter, the latter of which can be visible in broad daylight under certain conditions) and then it

AN ECLIPSE TO REMEMBER

would automatically point to, and stay centered on the selected object.

After the purchase of the LX200 mount package, the goal of observing the 2017 eclipse became more solidified. With the acquisition of the LX200 came a few smaller telescopes shortly after, partly because I'd became a telescope collector and secondly, I would later use those telescopes to establish a local astronomy club. By that time, I had also become a seasoned astronomer and quite an accomplished photographer of deep sky objects. It was at that point, that I set a goal to not only observe the eclipse of 2017, but to photograph it.

Later on, I would obtain a few smaller telescopes, including a Meade achromatic refractor. The latter would ultimately later adorn the main rig piggyback-style in a sort of telescope cannon, providing a dual-imaging setup on one solid mount. With two telescopes on a single mount, the LX200 became serious business.

STARTING TO GET SERIOUS

CHANGING FORTUNES

In October, 2014, my fortunes had begun to change in a positive way. After a couple years of being unemployed, I found myself accepting a job offer at a major hardware store chain. Making the news that week, was an upcoming partial eclipse which was predicted to occur in the late afternoon hours of October 23. I was worried that I would miss it, not just because of the weather forecast (which was looking gloomy, with an incoming rain storm), but also because that day would also be the day I would start my new job. And unlike the 1991 eclipse where just a mere 20 percent of the sun was blocked out by the moon, this eclipse – with more than 50 percent of the sun blocked out – would be what I call a "tease."

Fortunately for me, my first day was scheduled to be a morning shift, and with only four hours on the clock, I was to be off work with more than enough time to set up my telescope and camera. I clocked off work at noon, and shortly afterwards, I was making the half-hour drive home. Within minutes of getting home, I had my telescope and camera coupled up. The plan was to capture the eclipse with the ultimate goal of doing a complete time-lapse from start to finish.

Sadly, however, the time lapse never happened. That day, the forecasts from the week prior were now coming to fruition. Not only were the forecasts remarkably accurate, the rain and advancing storm clouds were actually more intense than forecast. The partial eclipse started at a little after 1:35 that afternoon, and it was raining when it began.

Despite periods of heavy rain and thick cloud cover, there were a couple quick breaks in the cloudy gloom and rain. It was because of those breaks in the rain, that I was able to capture two remarkable images of that partial eclipse. It was those images, and the experience of seeing it through my telescope (by way of my camera's viewfinder), that had me hooked on eclipses.

It would be a little less than three years until 2017, and by that time I was contracting a hard case of what some call "eclipse fever." One of the images I captured, was roughly 20 minutes after it started, and the other was perfectly timed at just before mid-eclipse. In that one, the moon was eclipsing around 48% of the sun's disc.

Not long after the eclipse, I would experience another change in fortune. A week later, I met a wonderful woman named Christine Reynolds. Immediately, we hit it off and after a few months of dating and a quick engagement, we were married. At last, I had a partner with whom I would share this wonderful celestial experience. In fact, I had talked about my plans of seeing the 2017 eclipse on one of our first dates, when I introduced her to my telescopes.

PLANNING FOR 2017

The better part of the next two years would be spent studying and preparing for the 2017 eclipse. In addition, a lot of planning would have to take place, as well as a lot of research.

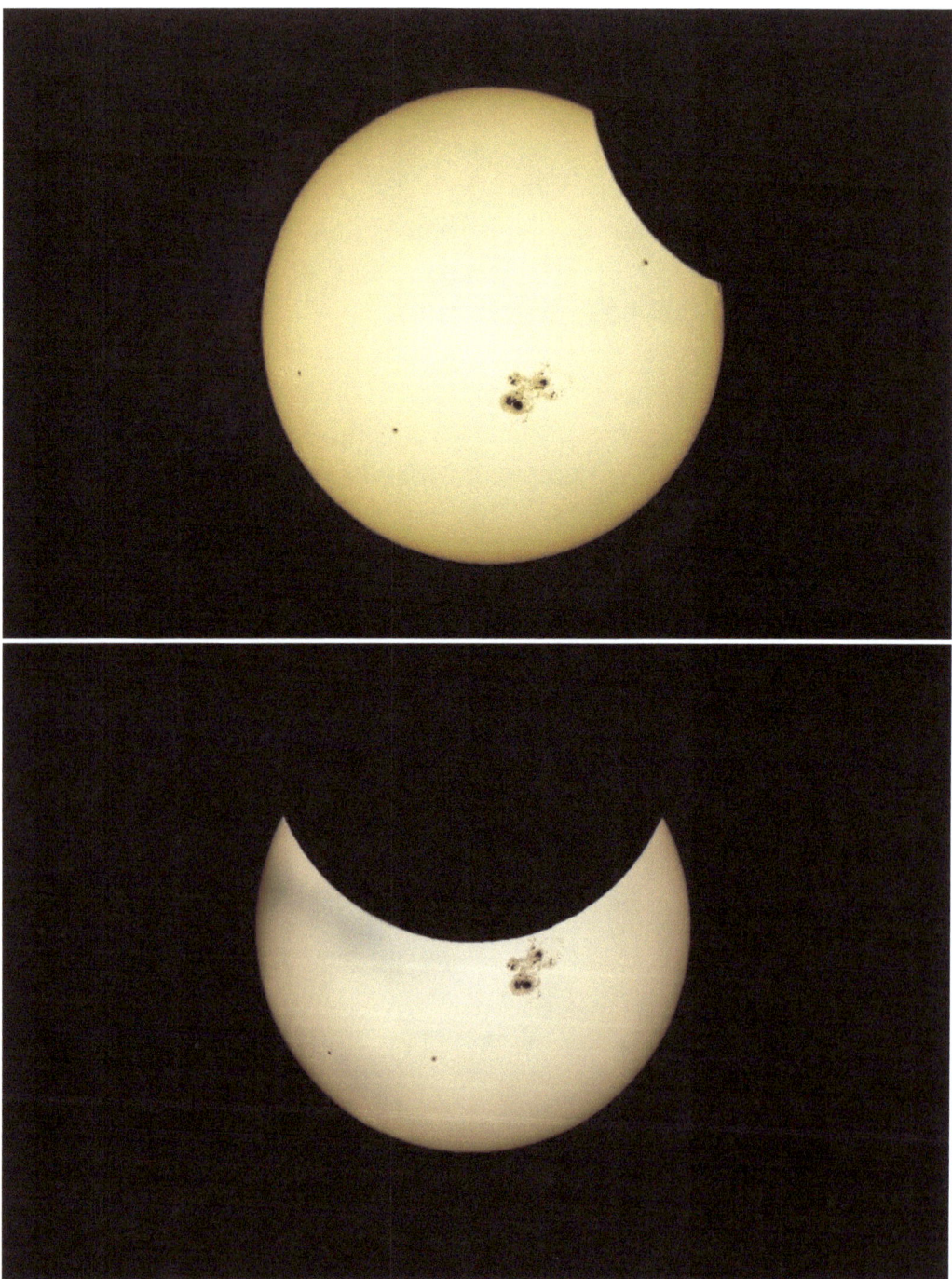

The October 23, 2014 Partial Solar Eclipse. The large dark 'spot' on the sun just below the eclipse was a large sunspot group, which was given the designation AR (Active Region) 2192. The two smaller spots below were, from left, AR 2195 and AR 2194.

As I began to research the 2017 eclipse, I stumbled upon a fantastic resource put together by a NASA astrophysicist named Fred Espenak. *Mr. Eclipse* as he was called, he was widely considered by many to be the expert in the field of total solar eclipses, and his websites *Mr. Eclipse* and *EclipseWise* provided an incredible database of eclipse prediction times. The latter also had an interactive map of the eclipse path, complete with eclipse durations based on observation site criteria. It was in one of his websites that I later discovered, that the 1979 eclipse and the 2017 eclipse shared some similarities. Among those similarities were that both eclipses would begin their path across the United States in the Pacific Northwest, and that some locations in Oregon would be witnessing totality twice in thirty-eight years.

SELECTING AN OBSERVING LOCATION

In its early stage of planning, selecting a spot to observe the eclipse was full of indecision. In early 2015, I began formulating a plan to observe the eclipse at some location within Oregon's scenic Willamette Valley region. The first and obvious choice was Oregon's capital city of Salem. It wasn't hard to choose Salem due to its relative ease of access off of Interstate 5, plus the fact that the neighboring community of Keizer would serve as a decent alternative. Another reason the Salem area was initially chosen, was because of the fact that one of my mom's cousins lives there and she was willing to provide us with amenities should we so choose. However, Salem *and* Keizer, Oregon each had several considerable disadvantages. The most notable disadvantage, was the fact that Oregon's capital city lay more than nine miles north of the centerline as far as the path of totality was concerned.

Eclipse viewing protocol dictates that if you're a first-time eclipse observer, the more totality one can experience the better the overall experience will be. In light of that fact, the nine-mile separation between Salem and eclipse centerline, meant that there would be a loss of eight seconds of totality (when you factor in lunar limb profile corrections and Salem's own elevation as a factor). The second disadvantage, was its location on Interstate 5. In early 2015, no eclipse forecast data was available in terms of projected traffic volumes. With an eclipse, it's widely known that thousands of people flock to see one, and since Salem is within reach of Portland and Vancouver (as well as cities like Albany and Eugene to the south) that meant major traffic jams were possible. I discovered that disadvantage firsthand in the summer of 2012. The third disadvantage was the sheer lack of public space. When you have a limited budget to work with, hotels or any form of lodging can quickly become undesirable alternatives.

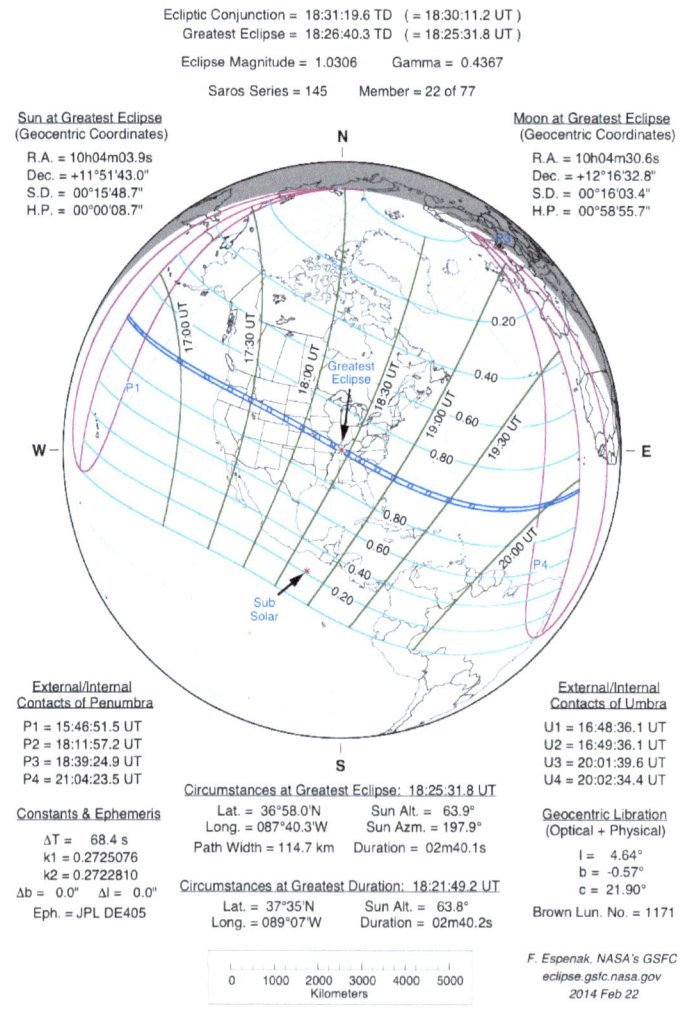

The eclipse prediction map for the August 21, 2017 eclipse. Contact times are listed in Universal Time.

From EclipseWise.com/Fred Espenak and NASA/GFC

AN ECLIPSE TO REMEMBER

After a few lengthy discussions and an exhausting study of Espenak's data, my wife and I ultimately selected the small Oregon town of Mill City as our initial observation spot. Nestled along Oregon's North Santiam River, Mill City is a small community situated 32 miles east of Salem on Highway 22. Eclipse predictions called for just over two minutes' worth of totality. It was also selected because it wouldn't be that far of a drive from Salem, and it was within a reasonable trip budget. I figured if we could swing a trip to Portland, Oregon for under $30, we could make Mill City work with a $60 round-trip budget. Another reason Mill City was selected, was because it was extremely close to the centerline. The centerline was only two-tenths of a mile north, and that meant that a distance such as that (less than say, two miles or so) would result in a negligible difference in terms of how much 'totality' we would experience.

A testing run in April, 2017, following the acquisition of a smaller Meade achromatic refractor (pictured here, mounted to our 10-inch Meade LX200). In this view, both Nikon cameras are hooked up to an HP commercial laptop running *DigicamControl*. Also seen is the white-light solar filter from Thousand Oaks Optical, which is a requirement for shooting the sun through a telescope.

A year and a half later, in mid-2016, most of the telescope gear that I would end up using, was finally complete and ready. Nights of deep sky object photography and a considerable amount of daytime solar photography would serve me well in terms of practice. It wasn't long though, that I began to think shooting it with one camera wasn't going to be enough. In August 2016 - with roughly one year to go until the eclipse, I had a sudden epiphany:

"What about photographing it with more than just one telescope? Why not photograph the eclipse with a dual setup?"

A plan was then formulated to shoot a couple HDR (High Dynamic Range) composites of the sun. From there, I would then combine images from both focal lengths to compile one of the sharpest eclipse images possible with amateur equipment. Finances at the time wouldn't allow the purchase of another ten-inch LX200 package, but not long afterwards I picked up a small and lightweight Meade refractor with a reasonably wide field of view (or FOV, for short). Best of all, it was acquired for a relatively cheap price. It had a field of view that with our camera would not only capture the inner corona, but some of the outer coronal streamers and any stars that might be visible. The eclipse was happening at a time of the year when the sun would be passing through the constellation Leo, and the brightest star in that constellation (Regulus) was only a couple degrees away from the eclipse. To accomplish this, it would be necessary to purchase another DSLR.

By that time, I had upgraded to Nikon's newest camera in the D5000-series line: The D5500. The upgrade was a requirement due to the fact that the D5100 I previously used, had reached the end of its useful lifespan. With more than 197,000 shutter actuations and an age of over five years old, my original D5100 began to fail. The D5500 was chosen most importantly because it was garnering excellent reviews online. Additionally, it had a sleeker, lighter body plus an innovative touch screen that swiveled out in a multitude of different angles. That touch screen alone would make the task of aiming and focusing the telescope, significantly easier.

However, a new D5500 in early 2016 was more than $600, and that cost was just for the body alone without any lenses. The cost of the D5500 kit I purchased was a monumental expense in itself, and I just couldn't justify the purchase of another "twin" camera. Considering those factors, the decision was ultimately made that another Nikon D5100 would have to be purchased. The reasons I chose another D5100 were obvious. One reason was the price. By 2016, the Nikon D5100 model was approaching six years old. By that age, most consumer-level DSLR camera bodies tend to become obsolete. With the age factor in mind, a relatively well-maintained D5100 body could be purchased for less than $300. Secondly, I had enough experience with the D5100, it seemed as though buying anything else was trivial.

Once the second camera was purchased, I began doing extensive testing on the entire system. Weeks of deep sky shooting that summer was often done with both cameras capturing in unison. This was done in order to test the computer's ability to command both cameras, as well as the ability of both cameras to work in unison without possibility of any glitches occurring. The fortunate similarity about both the D5100 and the newer D5500, is that they used the same wired shutter release cable plug and USB port, which eliminated compatibility issues with existing hardware and software.

By late 2016 most of the equipment was working without a hitch. Over a dozen test sessions were executed, with more than 25 deep sky objects and several sunspot groups, photographed that summer and fall. It seemed as though everything was going to plan, and everything would work like a charm.

Or so I thought.

THE EXCITEMENT BUILDS

With the eclipse six months away, it still felt like an eternity. By February, 2017, we were within the throes of one of the longest bouts of rainfall on record. However, despite the relentless rain, which seemed to have no end in sight (minus an exception of one break in January), feelings of hope and promise were building. At that time, I had everything I thought I needed. Two DSLRs, a pair of telescopes on one mount, and the laptop to control everything. In all of my testing and preparation for the eclipse, everything was performing as well or better than I thought.

In early testing of both camera setups, I thought I had a fairly adequate approach for solar eclipse photography, and that I was going to be capable of producing some outstanding eclipse composites. It was then, that I stumbled upon the works of Miloslav Druckmüller, who is one of the most accomplished eclipse photographers and eclipse photo processors in modern times. His work is simply incredible. In his work, Earthshine (the light reflecting off of the moon from a fully-sunlit Earth) shines brilliantly as well as the complete dynamic range of the solar corona. Even more impressive was his ability to capture the background stars that can often become visible during a total solar eclipse.

Up until then, my plan of action was to have the cameras manually controlled via external shutter releases and intervalometers, even though the eclipse totality phase would be executed completely manually – a tactic that goes completely against the wisdom and protocol of observing your first total solar eclipse.

(This was going to be my first eclipse, and the first major celestial event my wife would witness. Experienced eclipse experts recommend avoiding photography of an eclipse if you're watching one for the first time.)

I've never been one to follow conventional wisdom. And considering the lessons I would later learn, I would say that it's a pretty good thing I didn't.

In learning about Druckmüller's approach, I discovered that I had to abort the previous capture plans. The discovery of Druckmüller's work now meant I would have to switch to complete automation, meaning the computer would now have to control both cameras. That was especially true for capturing totality (even though ultimately, that would never happen, for reasons explained later). It also meant that I'd bring my wife into the fold. This wouldn't be just my event – it was also hers.

A JOINT AFFAIR

Now that my wife was on board, we also learned that we'd have to rethink our approach in terms of telescope optics. In the second full week of April, an announcement was made on Cloudy Nights (a forum for amateur astronomers) that Google and UC Berkeley were teaming up to create an eclipse feature to be called *Eclipse Megamovie*. The team leads at *Megamovie*, were in search of seasoned amateur astronomers and amateur photographers, who would be willing to contribute their photos of totality to the project. Eager to contribute to science as well as seeing the benefit of the projects goals, I joined *Megamovie.* A month and a half later, my wife joined.

The *Eclipse Megamovie* Official Photo Team Kit. Each member who participated in the testing and webinars received this kit as part of participating and contributing to the project

MORE GEAR UPGRADES

After my wife joined, we soon discovered that the Meade achromatic refractor which until then sat piggyback on top of the main telescope, was deemed inadequate. Its focal length and FOV were beyond the acceptable criteria sought by the *Megamovie* team, so we set out in search of a replacement. One of the smaller telescopes we have (a Meade ETX-70) was considered, and later rejected because it would have required heavy modification to the existing optical tube assembly. It would also mean we'd have to modify our existing setup and fabricate an entirely new mounting system for it, which was an effort that would have been cost-prohibitive. Towards the end of the search, our friend Jami generously loaned us a Meade Series 5000 APO (apochromatic) refractor, which not only met the qualifying criteria, it also lent itself to having a much sharper image on the camera sensor.

Our final eclipse setup, with the APO refractor mounted in place of the earlier achromatic unit. Photo taken June 2, 2017. The final weight of this setup (minus the 45-pound weight of the tripod itself) approached 90 pounds. Cameras were not installed at the time this photo was taken, due to much needed sensor cleaning.

Another thing that needed to be addressed was the need for a better system of solar filters. Up to that point, we had been using aluminized Mylar – which is basically the same material you find in emergency space blankets and Mylar party balloons – for a Rube Goldberg-inspired solar filter. The LX200's dust cap came with a circular hole cut into one side of it a little less than four inches in diameter, which I learned from its previous owner, was for aperture reduction while viewing the moon. For a filter, I would then cut a sheet of that Mylar into a square four inches across, and tape two layers of it over that hole with black gaffer's tape. While a successful means of filtering out most of the sunlight, it had a few disadvantages.

Every six months or so, we were replacing the filter material due to excessive wear, and the fact that aluminized Mylar doesn't really stand up to the test of time in a rough environment. Another deficiency in the use of that material is that the color of the sun was far too 'blue' in both the eyepiece and on the camera sensor. In conventional solar images, including the white-light continuum images from NASA's orbiting solar observatories (the Solar Dynamics Observatory and the Solar and Heliospheric Observatory, or SDO and SOHO, respectively), the image of the sun appears as an orange-yellow disc. In light of those factors, the bluish-white image of the aluminized Mylar seemed unnatural. It was also troublesome to process. Every image had to be de-saturated in Photoshop (which resulted in a loss of detail), then an orange mask was then applied which would simulate the look of the SDO/SOHO continuum images. After dealing with solar imaging in that manner for more than a couple years by that point, it was decided that for the 2017 eclipse, we would finally break open the piggybank and purchase a dedicated solar filter. The filter of choice was the long-trusted Type 2+ glass filter from Thousand Oaks Optical.

There was a small catch: Thousand Oaks Optical had recently discontinued that type of filter and by then, had announced a replacement product called 'SolarLite,' which is an impregnated polymer film with a lifespan of 25 years. I bought two: A 5-inch one and a 10-inch unit for my LX200. They would come in a polished, brushed aluminum filter cell and were supplied with a strip of self-adhesive felt tape for mounting. After receiving the new filters, we were quite pleasantly surprised at the tack-sharp quality and the amazing color production. The image at the eyepiece using both filters was a muted shade of orange with a hint of yellow. On the camera, the output was identical. We quickly discovered that these new filters were amazing, which meant that shooting the partial eclipse phases would produce outstanding images for a planned time lapse we had in mind. As it turned out, I was one of the first customers to buy the new SolarLite product. To this day, I swear by it.[1]

After rethinking our optical approach (and finally getting everything in place), I was stunned to learn that our planned eclipse site in Mill City, Oregon, wasn't all that ideal. Neither was anywhere else in the Willamette Valley. The first climate predictions for August began to be released about five months away from the eclipse. Those climate predictions were based on a historical average going back 15 years, in terms of weather outcomes on eclipse day. By mid-May, they were predicting more than a 50% chance of cloudy skies at totality for most of the Willamette Valley in Oregon, and in outlying areas including Mill City. This is due to the marine layer effect, in which low-lying clouds push in off the Pacific Ocean to areas further inland in the overnight hours, and then burn off later the next day. By mid-August, it is not uncommon for such a marine layer to burn off well past noon throughout much of the Willamette Valley (and even at home in Washington State's Puget Sound area, where it is more common). That means that at by the time the eclipse was crossing the Willamette Valley region, skies would have likely been clouded over at eclipse time. That meant that we had to change our plans.

PLANNING FOR MADRAS

Seeking a new location was relatively easy, and the search didn't take long. After posting on social media about our dilemma, a friend and fellow astronomer suggested we try Madras, Oregon as an ideal spot. The suggestion was made because the same predictions that called for over 50% cloud cover in Mill City, pegged Madras as being the best place to be – in the entire country no less – for eclipse viewing. The prediction called for a small chance of just 18% cloud cover. By eclipse time, cloud cover projections for Madras were almost crystal clear.

---

[1] As of this writing, more than four months have elapsed since the eclipse.

Another forecast factor was the average temperatures. By eclipse time, temperatures were forecast to be in the low 80-degree range. Under most conditions, a high temperature index near 80 degrees often coincides with crystal clear skies. We also chose Madras because of its scenery. A lot of roadside geology exists in the area, as well as stunning vistas of more than five Cascade Range volcanoes. The science geek in both of us was going into overdrive.

By the first week of June, we had settled on Madras, Oregon. A small city with the population of just 6,400, Madras is nestled within a basin in Oregon's central desert. The summer climate in Madras often features sunny skies and fantastic views. However, at first our initial plans included an overnight camp out in a field near Goldendale, Washington. We would then head down early the morning of the eclipse and set up. Goldendale was initially chosen due to a friend owning a 20-acre tract of land there. He uses it to host star parties for fellow astronomers, and since he was viewing the eclipse as well, he had been graciously willing to let us use it as a stop-over. It was also suggested because it would be roughly one and a half hours from there to Madras assuming normal drive conditions. At the time, we figured we could easily wake up a few hours before dawn and be down in Madras within an hour of eclipse time.

Then, the traffic and eclipse viewing population projections started coming in. and with that a necessary change in plans yet again. A couple weeks after Madras had been selected, the first traffic map projections had begun showing the potential for more than 1 million people filtering into the Madras area in the week leading to the eclipse. That would present a huge logistical challenge, especially if we were leaving Goldendale in the morning with any hopes of being down there and being completely set up by 9 a.m. Oregon's Department of Transportation (ODOT) had begun to warn travelers that in many areas, delays could last hours or even days. With those warnings, came further advisories that those delays could start not only the day of the eclipse, but last throughout the entire weekend leading up to it.

A brief plan materialized to spend the night before the eclipse down at the Cow Canyon Rest area on Highway 97, about 23 miles north of Madras. Those plans were soon called off when we called ODOT to confirm an internet rumor. The rumor was that the Oregon State Police would be monitoring rest areas for people camping illegally by way of setting up tents in rest areas, or by parking on the shoulders of major highways throughout the state. According to several ODOT officials, the rumors were largely true, so those plans were shelved since we would technically be considered 'illegal campers' at that point.

The rumors were ultimately confirmed, when the Oregon State Police sent out warnings on social media. Among those warnings, were the risk of being fined for watching the eclipse on the shoulder of any major highway within the eclipse path, or anywhere within Oregon's rest stops. In light of the warnings by the Oregon State Police and ODOT, we contacted the Oregon State Parks department. The call to Oregon State Parks was made upon hearing that they would open up additional sites for eclipse viewing. While they did confirm that they were opening campsites up for the eclipse, that idea quickly proved to be a bust. When they opened up reservations that month for 1,100 campsites in the path of totality, reservations booked within 11 minutes. Later on, when they opened up an additional 850 day-use spots that July, reservations booked almost instantly. Any hope of camping, it seemed, was doomed.

Time was quickly running out. After exhausting every avenue we could possibly travel down, with regards to accommodations, all hope was thought to be lost. Then, a coworker suggested something that became a miracle. The suggestion was to camp out on BLM land.

After some research, we discovered that a vast percentage of land in central Oregon are public lands. Large segments of those public lands are owned by the Federal government, and are managed by the Bureau of Land Management (BLM) and the United States Forest Service (USFS). The Madras area is chock full of BLM land, so we both perused the BLM website, and spent three days evaluating possible sites. The benefits to camping out on BLM land quickly became obvious. The first and foremost reason was that there were no fees or permits necessary to do so, a factor which lessened our financial burden of the trip quite significantly. The second is that there were no reservations required, and we could camp out as long as necessary. Before we had discovered the BLM land option, nearly every alternative we came across involved either a ridiculously expensive hotel or campsite reservation.

The BLM land option was literally one of the best miracles we encountered in our planning of the trip.

After looking at the overlay of the land (and various terrain maps), the decision was ultimately made that we would camp out on a tract of BLM land somewhere near the Madras airport, in a clearing just off of Pelton Dam Road where it turns into Elk Drive. We printed off at least a dozen pages of maps as reference material. The decision was made mainly because Google Maps (and Google Earth's 3D view) showed an unobstructed view of Mount Jefferson from that spot. It's often been said that if you have a view of distant mountains during an eclipse, it's quite easy to watch those mountains darken first, (that was one goal I had in mind – to see that darkening of Mount Jefferson). Eclipse projections had Mount Jefferson going completely dark about 20 seconds before our planned site would go dark. With the umbra moving across the land at 2,500 mph, it should also be easy to see it rushing towards us. It also looked ideal because there were no vertical obstructions that would get in the way of our planned observation of the eclipse itself.

By now, the testing runs for the *Eclipse Megamovie* project were concluding. As members of the team, we had to test our gear monthly, and attend monthly webinars that lasted anywhere between an hour to an hour and a half. The first of those webinars occurred on May 25th, the second on June 28th, and the last on August 3rd. Interspersed with those were group chat nights, where we were given the privilege of chatting with the folks in charge of the *Megamovie* project such as Vivian White and Brian Kruse at the Astronomy Society of the Pacific, Hugh Hudson at UC Berkeley (who created the project idea), as well as Dr. Laura Peticolas from Google. Being a part of the *Eclipse Megamovie* team also meant that we had to be dedicated to the project, and it also provided a fun way of getting to know other members of the project. Webinars were often held at around 6:30 p.m., and on two of those webinars we held test nights where we would use a quarter-phase moon as a stand-in for the eclipse. During one test session, we were actually able to use the sun.

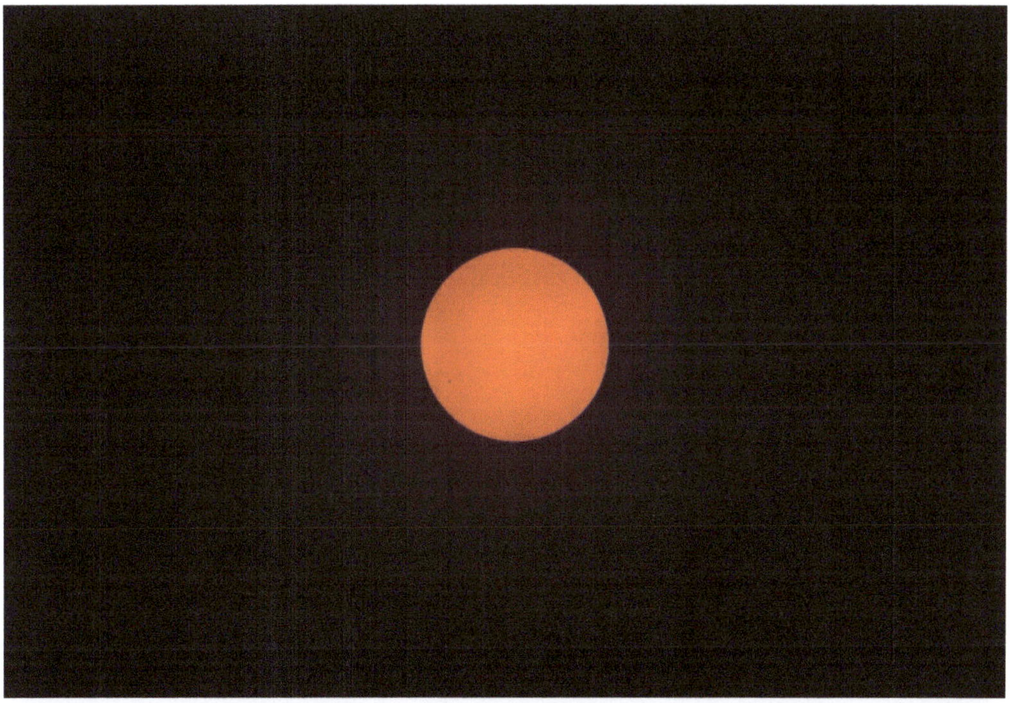

AN ECLIPSE TO REMEMBER

These two photos were taken as part of the final *Eclipse Megamovie* testing webinar, on August 3, 2017. Apparent in both photos, are the differences in focal lengths for both cameras, based upon the telescopes used. Top: Meade Series 5000 APO refractor/Nikon D5100; Bottom: Meade LX200/Nikon D5500

Testing would consist of taking a batch of photos with the camera setups and then we would submit those test photos to the *Eclipse Megamovie*'s upload server. One catch with photographing the eclipse for *Megamovie*, was that photos had to be geo-located through GPS. Cameras either had to have a built-in GPS (which would write the location of the photo to the image's EXIF data), or you had to use a combination of a geo-tagged smartphone photo and a file from a non-GPS-equipped DSLR. Since none of our cameras were capable of built-in GPS, we opted for the smartphone option.

The extensive testing with *Megamovie* also confirmed an earlier belief that everything would now have to be automated. A search of software programs to do the job settled on a free program called *DigicamControl*, which as it turned out was quite versatile. Even better, as a free program it also allowed for the control of multiple cameras.

By the end of the first week in August, things were finally beginning to fall into place. With two weeks to go until eclipse day, I was pleased to learn that my employer had graciously approved my request for time off. It had long been a goal, to have the entire weekend off beginning on Friday, since my scheduled days off fall on a Monday and Tuesday. Ideally, we would be down in Madras for five days. Those five days would include a day for travelling down, two days for setup and testing, followed by the main event on Monday and then a drive back either late Monday night or on Tuesday, the 22nd of August.

The next two weeks was spent waiting, and purchasing the last of the items we would need for the eclipse. Since we wouldn't be able to power the telescope from the 120V AC main power we have at home, we would have to control and power the gear remotely. That included buying a 12V deep-cycle marine battery, plus building a field battery to provide 12V DC to the telescope and cameras. The field battery consisted of the deep cycle battery nestled in a battery box, with five cigarette lighter plugs wired to it and a USB docking station (which would charge our mobile devices). Next, a homemade wiring harness was built consisting of power supply cables that converted 12VDC to 7VDC (for the cameras; this was done by way of a DC-to-DC converter), as well as shutter control and data transfer cables. We also had to figure a way to keep the radiant heat of the sun away from the laptop (which when heated by the sun, would overheat and fail to work) by applying a coat of white retroreflective spray paint on a large plastic storage tote. The same was done to a folding TV tray table, which would become a makeshift laptop desk out in the field.

FINAL PREPARATIONS

We also had to purchase several items not related to the direct mission of eclipse photography and eclipse observation, including supplies for our trip, and borrowing some camping equipment. After making those modifications and getting things in order, we had begun the process of clearing off more than 200 gigabytes of photo archives off of one of our laptops to make room for the mountains of incoming photo data. Since we were shooting with two cameras, a plan was also devised to do two subsequent time lapses.

One of those time lapses would be shot on the LX200, and the other, on the smaller APO refractor. The idea of shooting those time lapses and discussing them on social media caught the eye of a good friend and fellow photographer named Meg McDonald. Unlike our plans for Madras, she had ultimately decided on heading down to Albany, Oregon for the eclipse. After talking about time lapse editing techniques and formulating the requirements, it was decided that we'd collaborate on a joint time lapse video. The final plans for the time lapse and the joint collaboration would ultimately take place after the eclipse.

A week and a half before the eclipse, we were thrown a huge curveball. There was a storm front that had moved in, that clouded us out and nearly risked a total loss for the trip. Up to that point, the weather across the entire Pacific Northwest had been consistently hovering around 85-90 degrees with mostly sunny skies since late June. Almost overnight, the forecasts changed immediately and long-range forecasts for eclipse day were beginning to look downright gloomy.

In Portland, KATU's meteorologist Dave Salesky, began issuing forecasts that, for the Madras area, were bucking the previous climate predictions. Widespread rain was now being predicted, with Madras forecasts expecting torrential downpours on eclipse day. For the first time since June, my wife and

I thought about ditching the trip altogether. At worst, we selected a spot near Casper, Wyoming as a backup location because it was further away from Oregon and any risk of clouds that the storms would pose. At this point, it seemed like a bust.

Fortunately for us, though, the forecasts for Madras improved within a couple days of the first storm. What had been a forecast for rain and temperatures in the 60s for Central Oregon a week before the eclipse, had within a couple days turned into sunny skies and temperatures near 90. By midweek prior to the eclipse, the forecasts were looking exceptionally clear, with no risk of any inclement weather.

It was now Thursday, August 17th, with four days remaining until the main event. Temperatures in our hometown of Shelton, Washington were climbing back into the upper 70s and low 80s after a brief period of storms. Down in our planned site of Madras, temperatures were well into the 80s once again. A concern for us, was smoke haze from a fire that exploded on the west flank of Mount Jefferson a couple days prior. Satellite images showed the smoke column flattening out and drifting over Madras almost consistently.

By now, tensions and anxiety are running high. That day, I had been scheduled to work my job as a photo lab technician from 11:30 a.m. to 8:30 p.m. Reports on the news leading into the beginning of the week were holding firm on their earlier predictions of more than 1 million people filtering into the Madras, Oregon area. In fact, some of those predictions called for even more than that. One estimate had more than five million.

On the night of the 17th, a report from a TV station in Bend, Oregon, showed a traffic backup south of Madras that was more than 16 miles long, with wait times exceeding nine hours. The traffic backup was caused by more than 20,000 people heading to a large alternative lifestyle event scheduled near Prineville called Symbiosis, plus about 5,000 people heading to the Oregon Star Party being held in the Ococho National Forest.

Seeing the pictures and video, my wife called me at work in a panic, saying "Hey, we'd better head out now. Traffic backups are more than 15 miles long near Madras!"

AN ECLIPSE TO REMEMBER

After I got the call, I checked out the footage myself, then asked for a quick break to call the Oregon State Police and check Oregon's TripCheck website. At that point, I notified a supervisor of my need to be released from work early. My supervisor graciously allowed me the time off (my management team knew of my involvement in the *Megamovie* project and were amazingly supportive of it), so I clocked out at just before 7 p.m. and gassed up our SUV at a nearby gas station (plus filling a 5-gallon emergency gas can). From there, it was a brief stop at home to eat a light dinner, followed by packing of our rig. Packing our vehicle was done quickly, which included packing four telescopes and all of our camping gear. We finally left Shelton, Washington at 7:25 p.m. on August 17th. Our only planned stop was at a Walmart in Chehalis for food supplies and a last-minute purchase of a required piece of camping gear.

According to the GPS, Madras was four and a half hours away.

The GPS was about to be proven wrong.

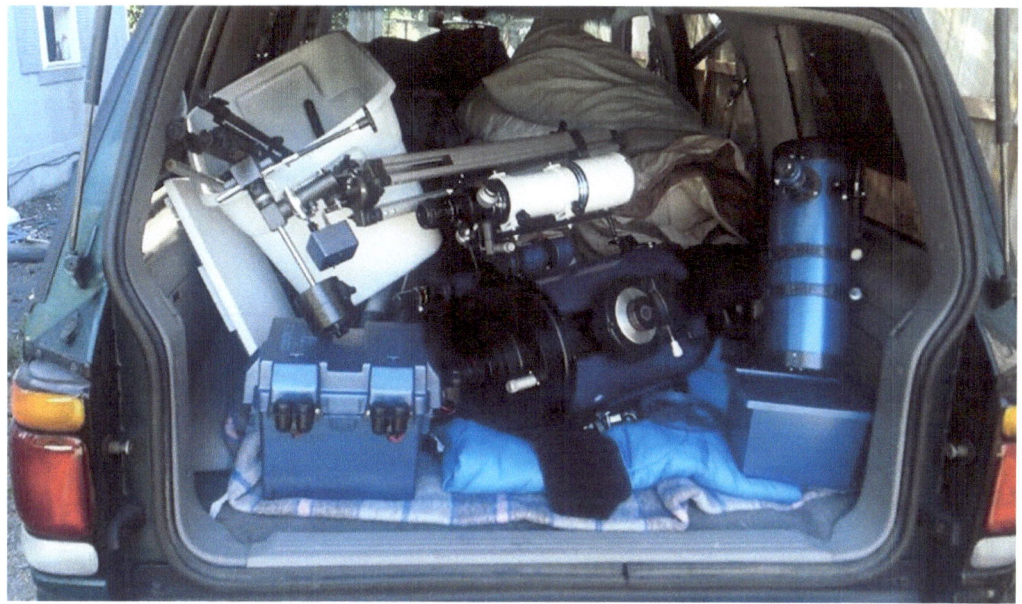

All packed up, and ready to go.

ARRIVING IN MADRAS

When we left Shelton, we had no idea what lie ahead. Our fears of what I call "eclipse traffic hell," were initially laid to rest with a preliminary check of the WSDOT mobile app which showed no gridlock from home to Portland. However, traffic maps according to Oregon's TripCheck were telling another story.

About an hour after we left, we made our planned stop at Walmart in Chehalis. Traffic between Shelton and Chehalis was smooth, with no obstructions or any form of gridlock. There, we purchased a couple of days' worth of food supplies and a lid for our portable toilet (which really was nothing more than a slightly modified five-gallon paint bucket). After we left, it was onward to Madras. As we were leaving, we decided that when we crossed over into Oregon we'd top off our tank in Gresham, then head over the pass at Mount Hood on US Highway 26.

AN ECLIPSE TO REMEMBER

An hour and fifty minutes after we left, we approached the community of Woodland, Washington. There, traffic came to an immediate standstill. As we were coming to a stop, I decided to check out the WSDOT mobile app. Through the app, we discovered that the backup was seven miles long, with delays lasting up to an hour. Our initial thought was that it was due to people heading down early for the eclipse, and that my leaving work early was justified.

The day before we left, we had equipped our SUV with two large *Eclipse Megamovie* banners announcing that we were a member of the official photo team (logos which were also on hats the team leads provided us), While being stuck in the gridlock, the other travelers sure gave our rig a look or two, likely because of the *Megamovie* banners.

As we soon discovered, our early assumption that the gridlock was solely due to eclipse travelers, was in fact not so much. We found out that instead of being caused only by eclipse traffic, the backup was due in large part by poorly-scheduled road construction (a fact we discovered by way of a few road warning signs a half mile into the backup). Apparently, the decision makers at WSDOT decided that eclipse weekend would be a fine weekend to schedule bridge deck replacement work on the southbound I-5 crossing over the Kalama River.

The roadwork reduced traffic to one lane causing a bottleneck scenario, stopping hordes of eclipse travelers heading south. Once we cleared that backup (which took roughly an hour and fifteen minutes alone), we then crossed into Oregon on Interstate 205. Immediately after entering Oregon, we exited I-205 and made our way east on Interstate 84. After about eight miles of driving we made our planned stop in Gresham. We topped off as planned, and purchased a couple snacks to soothe our hunger for the remaining two two-and-a-half-hour drive. As we were ascending the west side of US Highway 26, The view of Mount Hood's silhouette against the brightness of the starry night sky was a fantastic treat.

It wasn't long before we began descending down US Highway 26 on the east side of Mount Hood. Soon, we noticed a few wildfires burning off to the east near the Warm Springs Reservation. At first, they were noticeable as a faint orange glow off to the southeast, then as the further we traveled, we began to see the actual fires burning off in the distance as bright orange spots on distant hillsides. Because of those fires, we noticed that the air had a smoky aroma to it.

One remarkable sight that night was just how bright the stars were in the night sky. After you descend off of Mount Hood, and for about seventeen miles before the community of Warm Springs, Highway 26 follows a fairly straight path through a sparsely populated area (with a slight kink to the southeast about nine miles before Warm Springs itself). The few houses in that straight stretch, plus the fact that Madras doesn't have much light pollution itself meant that the Milky Way was extremely bright. The urge to set up for deep sky photography along the side of Highway 26 was quite high that night.

It was around 1:40 a.m., when we finally arrived in Maras. Despite running nearly two hours behind schedule, we were surprised to learn that we arrived quicker than what our GPS indicated based upon our departure from Gresham. When we first arrived, we were awestruck by the sheer size of SolarTown, a large grouping of campers positioned on a farm just east of the Madras airport. We estimated over 25,000 RVs and tent sites at that location. Additionally, as we approached downtown Madras, we were alarmed at the strong National Guard presence in the area.

After passing SolarTown, we began our descent into the basin in which most of downtown Madras resides. In the distance, we could make out SolarFest, which was a festival of around 10,000 additional eclipse watchers at the Jefferson County Fairgrounds situated on the southwest end of downtown Madras. Interestingly, as we arrived everything seemed strangely peaceful. We were the only moving vehicle on the road at that point.

Our first stop was at the Safeway just as you descend into Madras. We had initially assumed the store was open 24 hours, however upon walking up to the entrance we were greeted to locked doors and a parking lot attendant advising that they had just closed. After walking back to our SUV, we then headed to a gas station across town that was open.

AN ECLIPSE TO REMEMBER

There, we used the restroom facilities and purchased a couple soft drinks and a light snack to serve as an impromptu dinner. Afterward, we decided to hunker down and sleep overnight in our SUV back in the Safeway's parking lot. It wasn't long before we were asleep, and sleep wasn't comfortable considering we were sleeping in the front seats. After we awoke the next morning, our next stop was our planned eclipse observation site. First, we had to figure out how to get there.

STAKING OUT A SPOT

At 6:50 that Friday morning, we awoke. It was August 18, and the light of the rising sun was our alarm clock that morning. After waking up, we first decided to go on a small tour of the town and check out some of the sights. To our surprise, most of the businesses we encountered were open well ahead of normal business hours. After a short time of driving around, we spotted the offices of the Madras-Jefferson County Chamber of Commerce. Nestled inside that same building (and in the same lobby, no less), were also the offices of the Crooked River National Grasslands, and the United States Forest Service.

We arrived at 7:25 a.m., and quickly noted that the office wasn't open until 8. To our surprise, however, (and for what I think were obvious reasons, namely the upcoming eclipse weekend visitors) the front door of the building was unlocked, so we walked right in. Once inside, we had our perusal over dozens of maps of the area. Each map had greater detail than the ones I had printed online. Another – even

Descending down into Madras, Oregon on Friday, August 18, following our visit with the Jefferson County Sheriff. As this photo was taken, smoke haze from the Milli Fire near the Three Sisters volcanic chain began to fill the sky.

greater – surprise is that they were open for business.

The first to greet us as we walked in, was a lady working for the Madras-Jefferson County Chamber of Commerce. She was quite an informative lady, and in fact was also very helpful in giving us plenty of information about the eclipse and where to see the sights and sounds of the area. It wasn't long before she also advised us that the official in charge of the Crooked River National Grasslands (which also included all of the BLM land in the area), was on his way. As we were discussing our mission and the maps on display, more people began to filter in, including quite a few people from out of the country.

"Hi, where are you guys from?" we asked.

"We're from the Netherlands!", one couple replied, adding that they flew in to Portland International Airport at 4 a.m., and had just arrived in Madras. Another couple we noticed, had a thick accent from the United Kingdom. "This is our second eclipse," the couple said in unison, saying they had previously witnessed the 1998 eclipse while on a cruise in the Caribbean.

Both couples were quite talkative and a pleasure to hold conversation with, and I then demonstrated the dynamics of the eclipse using a tool called the "Yardstick Eclipse" (explained in later detail), as well as basic solar system science. By that time, more than 10 people occupied the small lobby, and as we began discussing the maps and details of the eclipse with one of the couples, the US Forest Service official in charge of the area had finally arrived. After a fifteen-minute wait, we then had an hour-long discussion with him about the specifics of where our planned site was, and how we would get there.

After discussing our plan, we were given a half dozen maps of the area which included a large fold-out map of the Crooked River National Grasslands, and Ochoco National Forest areas to keep. It was around 9:30 a.m., and one thing that we really enjoyed about the trip up to this point, was that everybody we encountered was rather helpful. For us, it seemed that everyone loved our presence down there. Being 'out-of-towners' as it were, it was as though we were treated like royalty.

AN ECLIPSE TO REMEMBER

After we left, we headed out and within 40 minutes we were scouting our first planned observation spot. It was the location we had scouted a couple weeks before on earlier maps, where Pelton Dam Road ascends uphill to become Elk Drive. At the top of the hill, there are a few dirt roads that head west off of Elk Drive. One of those roads forked out to the north, out onto an overlook of the Deschutes River.

Arriving at the first location was quite easy, despite nearly getting momentarily lost. At first, we drove out west of town on Belmont Lane for about six miles, until we reached a dirt road on the left marked SW Cinderpit Lane. The road looked rough and unfavorable, so we backtracked for a half mile until we turned left onto Elk Drive. There, we continued north for approximately three miles, and made another left onto an unmarked road. At the end of that road was an overlook with a stunning panoramic vista of Mount Hood, Mount Jefferson, and the Deschutes River Canyon.

Mount Hood, photographed from our first planned site. Immediately after this photo was taken, we would be forced out of this location.

It was here on that overlook, that we stopped. The map indicated we were on Forest Road 66330011. As we were arriving, we noticed a few wild coyote prancing across our path. The grass here, was extremely high, and extremely dry. The high grass, coupled with our thermometer indicating that it was already 85 degrees, had us nervous. It was feared that our vehicle's exhaust would ignite the grass and spark a fire, which fortunately for us never happened.

GETTING CHASED OUT

After taking in the scenery (and taking a few photos of both Mount Hood and Mount Jefferson), we stopped short of a Y-shaped junction in the road, and proceeded to set up our site in a cleared-out spot. As we began the process of setting up our tent, the situation then grew serious.

In the distance, a large dark-colored Dodge pickup truck was rapidly approaching our site. At first, we thought they were arriving to watch the eclipse. It didn't take long however, for the driver to get confrontational.

As they came to a stop, the driver threatened us with physical force if we didn't leave what they claimed was "their property."

"You guys are going to have to leave or we'll have no choice but to call the cops!" they exclaimed, adding "You guys are on private property and have no permission to be here. If you don't leave, things will get physical!"

"Uh, you'll have to excuse us, but the maps say otherwise," we insisted. "In fact, the GPS and the Avenza Maps® **app** (provided by the USFS for displaying BLM maps) on our mobile devices both say we have permission to be here."

"No, this is private property! The maps are wrong!" they responded, adding that the GPS couldn't possibly be correct despite the display readings clearly stating otherwise. They further insisted that the maps were wrong, by proclaiming that there were signs posted on Elk Drive warning against trespassing, saying "You guys clearly didn't read the signs back there! They specifically said 'No Trespassing' and that means get off our land!"

After several moments of intense debate, an argument broke out between us and the two in the truck. It seemed all efforts at trying to prove them wrong were being met with deaf ears (and blind eyes, it seemed) so my wife and I decided that to avoid the possibility of physical conflict, it was best if we left.

After we departed our first location, we headed to a huge campground set up near the Madras Airport, called SolarTown. At the time, well over 20,000 were camped here (a figure that would later climb to well over 100,000).

Looking east at monumental gridlock on NW Dogwood Lane. The line of stopped eclipse travelers heading into SolarTown extended three miles to the east, ending at the junction with U.S. Highway 97.

We took a detour here because – unbeknownst to us – one of my wife's cousins was camping there. It took us quite a while to locate him, where we spent about an hour and a half relaxing before we had to take off again.

Before we left SolarTown, we took in a few hot air balloon demonstrations that were taking place, as well as a few photos of Mount Jefferson and the surrounding scenery. Afterward, we decided it was best to rest up, as we were fairly exhausted by that point.

At SolarTown, we discovered a free day parking area which had been set up for travelers as an impromptu rest stop. It was also a spot where travelers could come in on the morning of the eclipse and watch the show, without paying for a campsite. It was a point of relaxation for us, and provided much-needed relief from all the driving we had done. The amount of liquids we were also consuming to replenish our thirst (and to keep us from dehydrating in the scorching heat) meant that we had to use the restroom quite a bit.

Fortunately for us, SolarTown had a row of about 20 portable toilets, plus a portable shower. Following a lengthy nap, we left SolarTown in search of another spot. Our exit from that site was an unenjoyable experience, due in large part because hordes of travelers and last-minute arrivals were now beginning to arrive in large numbers, creating a localized backup that clogged every available side road for at least two miles. On NW Dogwood Lane (where the entrance to SolarTown was set up), the backup was so extreme that at one point, a couple deputies from the Jefferson County Sheriff's Office were sent by motorcycle to help disperse and control traffic. When we came across one deputy, we asked where some ideal spots on public land were, and how to get there.

"Excuse us for a moment," we said, getting his attention as we drove by.

"How can I help?" the deputy responded.

"We're with the *Eclipse Megamovie* project and we're in search of some public land. It seems as though we're striking out, can you direct us to a good spot?" we asked.

"Sure!", he quickly replied. "Head back into town until you come to 'B Street' and make a left," he said. "After a few miles, B Street turns into Ashwood Road and you'll see an entrance to the prison on the left. Directly across from that, is an unmarked dirt road. Make a right on that road and you're there!"

After leaving, we then headed back down to the Crooked River National Grasslands office, where we alerted the officials about what happened to us at our first site. Then, we alerted the Jefferson County Sheriff's Office, where we would surprisingly learn that it wasn't the first time that couple ran people off without the authority or legal rights to do so.

"That's not the first time we've had reports of those guys doing that!" exclaimed one deputy at the Jefferson County Sheriff's Office headquarters.

"In fact, we gave them a warning yesterday for it!" added the office receptionist, saying that they've been hearing reports of it happening all across the Madras area. Also comforting to know, was that the maps were correct, which meant that we weren't wrong in being there. Hearing the assurance that it wasn't the first time it happened, also made us feel a little better. However, for our peace-of-mind and our sanity we decided it was best to avoid that area.

Following a discussion of about 45 minutes about possible alternative sites, the official we spoke to earlier, had advised us about a spot just south of the Deer Creek Correctional Facility, just three miles east. Ironically, it was the same location we were advised about as we left SolarTown. In fact, the directions he gave us were almost verbatim compared to the directions we received from the local sheriff's deputy we encountered earlier.

To get there, we had to drive east on B-Street as indicated, until it turned into Ashwood Road. At that point, Ashwood Road descended for a couple miles down a winding path into the bottom of a shallow valley. Before long, we spotted a gated entrance that led into a paved road, that continued north to the correctional facility. Directly across from the gate, was the entrance of the dirt road that led to the spot in which we were directed. At the entrance to that road, stood a huge sign posted by the Forest Service which welcomed eclipse visitors.

One thing we immediately noted, was that the road was rough, rutted out, and extremely dusty. Along both sides, were people in makeshift campsites, campers, and even one man sleeping in his car. After traveling for just short of a mile, we came to a rather large clearing off to the right side of the road about the size of a Walmart parking lot. When we arrived, we spotted two other campers nearby. Immediately, we began setting up within this clearing. Within an hour, we had our tent set up, our portable loo, and our camp chairs. It was now 4:30 in the afternoon, and we both agreed that it was time for a nap.

Sleeping in the tent was difficult at first. The nap didn't last long, partly because of the strange location we were in, and the fact that it was unbearably hot that afternoon. Secondly, both my wife and I had an unsettling fear of being attacked by wildlife, including wild coyote and the dreaded Western Red Rattlesnake (fears which we would later realize to be unfounded). However, during that first full night there, it made for an extremely restless sleep. Making it even harder for us to nap, was the ominous presence of a large dark-colored bull about 550 feet away. The term 'ominous' is used here, because it seemed he wasn't too happy about our presence, either.

After a couple hours, the heat from the late afternoon sun woke us up from our nap. After a quick lunch, we then relaxed out on our camp chairs. It was a peaceful afternoon, as we watched the bull graze lazily on grass and shrubs. What made it lonely for us at that spot, was that none of the other campers there, came over to visit us.

Smoke haze from the Warm Springs Complex fire fills the horizon, seen in this shot from our second location on the night of August 18, 2017.

After making dinner at around 9:30 – which consisted of a few peanut-butter-and-jelly sandwiches and a bag of chips, we turned in. That night, as we were preparing to sleep, we noticed the sunset was particularly striking. A sun pillar the shade of vermillion red was lighting up the bottom of some lenticular clouds that had formed over Mount Jefferson. To our north, the haze and smoke from the fires near the Warm Springs Reservation was still going strong from the night before and some hot spots were still visible, glowing off in the distance.

After falling asleep, we were soon roused out of bed at around 1 a.m. by the sound of a large truck making a ruckus on the road next to us. We first thought we were under attack again by the couple from our first spot, then I realized that the truck in this situation was not a Dodge. It had extremely bright LED off road lights, that were so bright they lit up our site (and the inside of our tent) from a half mile away. Once it got to where our tent was, it stopped and then drove off. Before long, we were falling back to sleep, which was restless that night. The next morning, we would set up our telescopes, begin final equipment testing, and then wait until show time on Monday.

Come tomorrow, nothing would go as planned.

AN ANGRY BULL

The next morning, we awoke shortly before 9 o'clock. While making our breakfast, we immediately discovered that the ice in our cooler had mostly melted. A small fist-sized 'iceberg' and a lot of cold water, was all that remained of what was once three large 20lb bags of ice. Taking note of our dilemma, we decided to head back to town to grab more, with the eventual plan of coming back out to that site.

There was a catch, though: Both of us don't like the thought of leaving a tent unattended while running an errand, out of a fear of it being stolen or ransacked. With that fear at the back of our mind, it was decided that we would break our tent down (and deflate our air mattress) so we could head back. As I was deflating our air mattress, the sound of the air pump's deflating action started making a screaming noise similar to that of an injured dog. Off in the distance, the bull we saw the day before had been startled by the noise. As I kept pumping the air out, the bull began to walk with a brisk pace toward our site. Sensing imminent danger, I began to work more quickly. While I'm doing so, I had noticed that it began charging towards us. My wife, who was calling her son in Olympia, was startled when I alerted her to the danger we were facing.

"Hey love, I think this bull is charging toward us!" I shouted, in a sense of urgency. "We need to hurry with this tent!"

"Can it wait?" my wife replied, "I'm about to call Shawn."

"Look out your window!" I exclaimed, as the bull was closing in, "We have a 2,000-pound bull that's literally charging us!"

"Oh CRAP!", said my wife, who then immediately tossed her phone on the dash and made a mad scramble with me to pack up our site.

The bull that charged our tent site is seen in this photo taken moments before we began dismantling our tent. This photo was taken from a distance of approximately 750 feet.

When it got to about 150 feet from our SUV, we nearly began to panic. When I finally grabbed her attention, we both then shoved the tent, partially-deflated air mattress, and the tent poles (which weren't completely folded up) into the back of our SUV. After making sure everything was grabbed, we ran for it. Almost immediately, it was decided that we were *not* heading back to that spot ever again. Disappointed, we realized that we were now 0-2 for ideal sites.

About 25 minutes later, we arrived back at Safeway. While grabbing ice, we also picked up a couple more boxes of granola bars, plus a case of energy bars and Gatorade. While we were at Safeway, we noticed an oddly-modified Chevrolet Caprice Classic sedan in the parking lot. It had been modified in a "Zombie Apocalypse" kind of way, we noticed, and I snapped off a photo of its backside (a camper that parked in front of it immediately afterward blocked my chance at getting the front of it).

Judging by the looks of this car, a zombie apocalypse was about to begin!

AN ECLIPSE TO REMEMBER

Seeing the superstition which surrounds total eclipses, it was no surprise that we saw quite a few cars with similar modifications during our stay in Madras. After leaving Safeway, we quickly headed back down to the office, for what would be our third and final visit. By then, it was around 1 o'clock in the afternoon.

Before we left, and earlier the night before, both of us had scouted out a spot which looked to be on a ridge just north of the correctional facility we were just south of. Our goal on that last visit was to find it, since it looked like it would provide an unobstructed view of the horizon.

The fold-out map, my GPS, and the Avenza app (which I downloaded at the suggestion of the USFS official there), all concluded that it was on a tract of BLM land. The key was getting there, so we asked the Forest Service rangers there for directions.

"We are having difficulty finding an ideal site! We just got ran off by a wild bull at the last site! Can you help us find this spot?" we exclaimed, pointing to the map.

"That spot is fairly easy to get to," the Forest Service ranger stated, before giving more specific directions. "Go north out of town until you hit the intersection with Highway 97. Then head north on 97 for a half mile until you reach NE Loucks Road—" (pronounced similarly to the word "loud"). "Take a right on that, and follow it out for three and a half miles. In the last mile or so, the road makes a couple sharp zig-zags, then straightens out."

"Then what?" both of us asked.

"After it straightens out," the ranger responded, "you'll come across a cattle gate across the road. Just after the gate is a narrow dirt road on your right. Take that, and follow it out along the base of that ridge. There'll be two roads on the left. Either one of those two take you up to the top of it."

"Thank you so much for your help!" we said. "You guys have been absolutely fantastic!"

"We need to warn you," the ranger advised us, adding "The first road is very steep and judging from the looks of your SUV, you might bottom out at the top of it."

After discussing what had happened to us with the bull, and our previous set of experiences, we left for our third and final attempt. As it turned out, the directions we were given, made it extremely easy to find.

Shortly after arriving at our final site. The view looks west, with Mount Jefferson just over the hood of our SUV.

At around 2:30 p.m. on Saturday afternoon, August 19, we arrived at the top of that ridge, in what appeared to either be the remnants of an old cinder cone crater or – as we'd ultimately learn – an old abandoned gravel pit. It was here, at 44°38'51.42"N, 121° 2'48.10"W, that we'd found our "home" until the eclipse's final act at fourth contact.

When we arrived, we were pleased to find a couple from Canada, who had driven down from Vancouver plus a young individual from Abbotsford. The presence of those few people on that ridge eased our concerns, especially in the aftermath of the events that took place the day before.

The former had arrived in a custom-modified Volvo tractor which was converted to an RV. It was equipped with enough reserve power to last an eternity. The tires alone on it were nearly as large as military troop transport tires, and there were two full-size spares hoisted up on a platform similar to those seen on delivery trucks.

We struck a conversation with them, and were invited for coffee and a mini potluck lunch. The younger man who was parked about 500 feet away, had also came up to our perch on the ridge, and invited us to use his hammock he had perched on the lone evergreen at the top of the ridge. Both the younger man and the older couple frequently came up to where we were and chatted with us, which made the experience of being on that ridge less lonely and more of an enjoyable one.

When we arrived, I was also elated to learn that this location had by far, the best view of any place we'd been at, previously. Our location the night before, was down in the bottom of a shallow valley with not much of a view of anything except the sky. Mount Hood barely peeked above a nearby ridge, and the lights of the correctional facility glared upon us at night.

The previous spot we were at, also wasn't nearly ideal for trying to capture the entire experience. The spot we were initially chased out of on our first full day, was also less than ideal, especially with the power lines that crossed the sky. The view of the Cascade crest at this new spot, however, including Mount Jefferson, Mount Hood, the Three Sisters and points in between was totally unobstructed.

With the view being completely unobstructed, that meant that not only would we see the eclipse's show in its entirety, we'd also see other aspects of the eclipse, including some that most don't usually see. As we arrived, we were surprised to see the large amounts of rock, which made you feel like you were in Arizona's rocky desert. To our surprise, we soon noticed some of the rocks at our site contained fossilized plant segments. On top of seeing the rocks, we were stunned by the sight of scattered shotgun shell casings.

Another surprise was that we could see where the centerline of the eclipse's "path of totality" was, with not only binoculars but with our unaided eye. At our location, we were a little less than four miles south of it. From our viewpoint, the centerline passed through the middle of a couple large fields and bisected one barn in half. Contrary to popular eclipse viewing protocol which always follows the rule "go to the centerline," I chose the spot because it had outstanding potential to see Baily's Beads (the last few specs of sunlight peeking out from behind the moon) in their finest display. Since the moon's umbra was more than 65 miles wide, and being four miles from the centerline, there was also a negligible difference in how long totality lasted with that difference in mind.

Upon our arrival, we immediately set up camp, ate lunch, and then took in the scenery for about an hour. As previously mentioned, the view from this location was outstanding, which afforded an excellent opportunity to do some photography of Mount Hood off in the distance, and Mount Jefferson. One thing we saw in addition to the already thick haze from wildfire smoke near the Three Sisters (from a fire in which officials named the Milli Fire), was a new smoke plume about five miles away off to our west. At first glance, it seemed a new wildfire had sprouted up, then soon we learned the tragic origins of all that smoke. While we were perusing Facebook, a breaking news alert had been made by Seattle's KOMO-TV 4. In the linked article, it mentioned that a plane had crashed while on approach to the Madras Airport, which itself had been hosting an eclipse 'fly-in' event.

After we got our tent ready and after having taken in the scenery, we then napped for a couple hours. One thing was certain: The dry heat of the high Oregon desert and the fact that we weren't used to it, being from the "wet"

AN ECLIPSE TO REMEMBER

side of Washington state, made fatigue a real factor for both of us.

It also made us sleep a lot despite being adequately hydrated and nourished. After sleeping for what seemed like a few hours, we awoke from our nap around two hours before sunset. The skies that night were full of high clouds that were drifting by, and it was those high clouds that made for one of the most impressive sunsets we both had ever witnessed. The shades of vermillion and deep shades of orange were forever etched into our memory.

It was that night that we would also set up our gear. First came the task of setting up the 50-pound tripod for the Meade LX200. Setting it up was somewhat difficult without the aid of a compass or bubble level. In the process of unpacking the imaging gear and all of the telescopes, I realized I had made the mistake of leaving the leveling compass at home. The leveling compass is a compass that sits into the recess of the top leg adjustment lock knob, and has a built-in bubble level for ease of alignment. After what seemed like an hour of frustration, I had an epiphany:

The cell phone reception for both of us was quite surprisingly strong for us on that ridgetop. We were consistently achieving four bars of 4G LTE service while there that entire weekend (despite predictions that cell phone reception would fail), and with that in mind I downloaded a few compass apps plus a bubble level app on my Android-powered smartphone. That turned a frustratingly difficult setup into an easy one.

The only thing left at that point would be leveling out the tripod using one of the bubble level apps, plus adjusting the latitude adjustment of the equatorial mount to 44 and a half degrees, a three-degree adjustment from our home latitude of 47.2 degrees north. Then came the monumental task of lifting the 95-pound LX200 setup onto the tripod, where I would leave it alone for the night. At nightfall, I would refine the alignment of the telescope even further. Refining the alignment of the telescope took roughly 45 minutes, which included a rough-in alignment on Polaris (the "North Star"), and a series of 15-minute drift-alignment procedures to ensure tracking accuracy.

TOP: Setting up the gear for initial preparation.

BOTTOM: A fiery sunset announces the end of the day over Madras, in this photo taken Saturday, August 19, 2017.

AN ECLIPSE TO REMEMBER

At our final location, taken after setting up the telescopes. Note the *Eclipse Megamovie* banner on the back windows of our SUV. The view looks roughly south. At this time, the telescopes were not polar aligned. Polar alignment would be done at nightfall.

While setting up, we had noticed that one of the fires burning near the Three Sisters/Mount Bachelor area had flared up, casting a reddish glow off in the distance. We weren't too concerned about the fire itself as it was more than 30 miles away, however we were somewhat concerned about the fire's smoke column, which was getting thicker as night fell. Both of us were also surprised to see the fires near Warm Springs had seemingly extinguished. Amazingly (and for a good reason), we never saw smoke from those fires for the rest of the weekend.

As we were preparing to sleep for the night, we noticed headlights from a pickup hauling a 30-foot RV trailer coming up to the top of the ridge where we were perched. It later turned out to be a family from California, who had traveled 14 hours from the Bay Area to see the eclipse. It was well past nightfall, and I had assumed that they didn't see us, so we then went straight to sleep. Sleep that night was restless, and anxiety ran high on that first night up there. A couple times during the night, we were awakened by the sound of something rustling around near our tent,. Both times, we quickly found that it was nothing more than a few stray dogs in search of food.

"WHO'S THE GOOGLE GUY?"

It is now Sunday, August 20, 2017, and the eclipse is just one day away.

That morning, we were awakened at 8:30 by the unmistakable sound of a Bell JetRanger helicopter flying overhead. To our surprise, it came to a hover directly over our ridge. A quick glimpse with the binoculars showed it to be an unmarked news helicopter from a Portland, Oregon television station. As it hovered overhead, our Facebook notifications began going off.

Among the notifications, was an alert. The ABC affiliate in Portland, Oregon. KATU Channel 2 had gone live over Madras for their Sunday morning newscast, and were showing a live view of a makeshift campground on a farm about a mile and a half west of us down in the valley that lay between us and downtown Madras. At that campground were over 1,000 campers. As we woke up and prepared for the day, we noted that the previous night's anxiety faded to eager anticipation for what lie ahead. The smoke column from the nearby Milli Fire near the Three Sisters had also seemingly settled down, at least for now. For the first time, we were able to lay eyes on the North and

AN ECLIPSE TO REMEMBER

Middle sister. Seeing that, we took advantage of the opportunity and took a photo of them.

A quick check of my phone showed promise as well. Three friends of ours had expressed interest in joining us for the eclipse. One was from California, and the other was from up near Snohomish, Washington. The third was a fellow high school friend of mine. Up to that point, we had felt somewhat isolated on our ridgetop vista.

With only a few people up there, who largely kept to themselves, things for both of us felt a bit lonely at times. A little later on, the couple from Canada had invited us down for coffee and a quick breakfast, which we happily obliged. They also gave us a tour of their RV. Then, it was onto a quick run of testing for the eclipse, where I took a few test shots of the sun with both telescopes.

It was there, during my first testing run, that we had experienced a sort-of celebrity status on that ridge. While I was setting up and testing the telescope's tracking accuracy (and aligning the optics of both telescopes, which ended up being misaligned during our travel down), the family from California had come up to chat with us. The imposing sight of two large telescopes were the first thing that caught their eye.

Even from a distance, a ten-inch Meade LX200 with a moderately-sized refractor piggybacked to it looks quite impressive. It gives the appearance that some serious astronomy business is about to happen.

Upon reaching us, their first words were "Who's the Google guy?" Even though it was sort of anticipated, that question alone had both of us startled. At the time, we also wished that it wasn't asked of us, but it was a price we paid for our participation in the *Megamovie* project.

EXPLAINING THE MEGAMOVIE PROJECT

The *Eclipse Megamovie* project banners. These were done for our vehicle to advertise the project and to foster interest in it, as well as the eclipse.

(Author's note: The Google, UC Berkeley and Astronomical Society of the Pacific logos were at the time of the creation of the banner and as of this writing, sourced from public domain files. They are used here under Fair Use, and no copyright infringement is intended.)

As it turned out, the *Eclipse Megamovie* banners which adorned our vehicle, turned out to be quite a conversation starter while on that ridge. Nearly everyone asked about it. The first thing that was necessary, was to explain that we weren't officially from Google (or UC Berkeley, or the Astronomy Society of the Pacific, the logos of which also adorned the banners as project sponsors). The second thing that was necessary, was to explain exactly what the project was. The *Eclipse Megamovie*'s goal was a mission of citizen science, with the intent to capture 94 minutes' worth of unimpeded totality.

While the eclipse was no more than 2 minutes, 40 seconds at its longest (in Hopkinsville, Kentucky), the entire duration of the eclipse was 94 minutes. That's how long it would take for the moon's umbra to travel from Lincoln City in Oregon to just south of Cape Island, in South Carolina. The mission of the project was to station as many photographers as possible along the path of the eclipse for the sole purpose of capturing totality. The resulting photos would then be algorithmically stitched into a 94-minute time lapse. Following that, astronomers would use that data to study small changes in the inner solar corona during the course of those 94 minutes.

Some 1,400 photographers volunteered for the project. Being a member of the *Eclipse Megamovie* team also gave us a bit of a prestigious honor (and an esteemed privilege), and we were given a hefty dose of *Megamovie* gear for compensation and our participation in the project.

Among the gear we got, were two baseball caps with the *Megamovie* logo, a lapel pin, around 80 pair of *Megamovie*-branded eclipse viewing glasses (to hand out to others), plus a "Yardstick Eclipse" tool which helped immensely with selling the benefits of the project and with demonstrating the science of a total solar eclipse. The Yardstick Eclipse was a folding yardstick with two wooden beads (scaled in size, Earth was a one-inch ball, and the moon was a bead ¼" in diameter) and a couple supports for both.

A pre-eclipse setup of the Yardstick eclipse, illustrating the entire scene. In this photo, it has been set up on top of the telescope rig. At the time this photo was captured, the telescope was aligned on the sun.

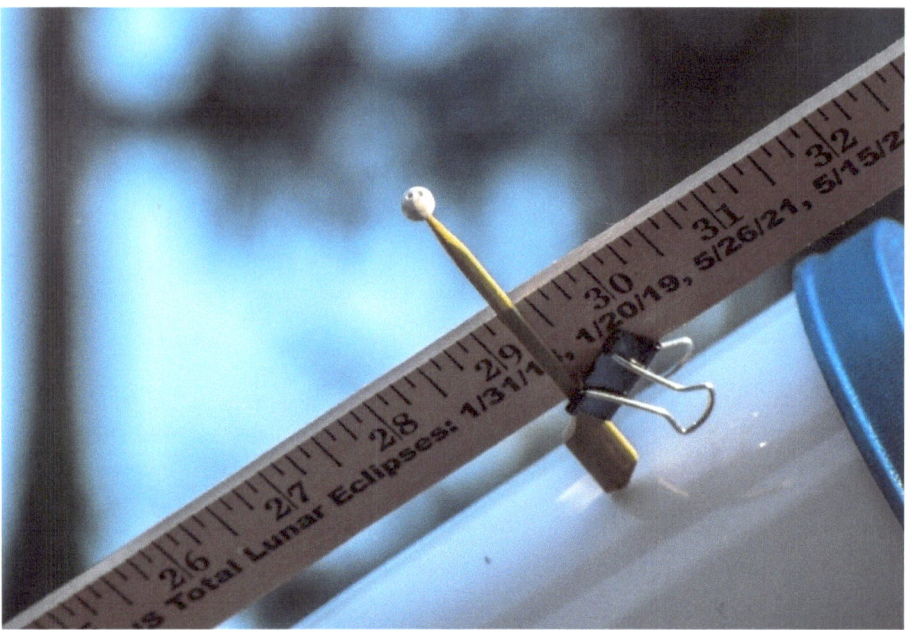

A pre-eclipse demonstration of the Yardstick Eclipse tool from the *Eclipse Megamovie* project, illustrating how it works. In the last photo, the darkened spot in middle of the 1-inch ball representing Earth, is the umbra of the micro-scale moon. Photos taken August 3, 2017.

To demonstrate an eclipse, you would place the 1-inch ball representing Earth at the 1-inch mark, and the ¼-inch diameter bead representing the moon at the 31-inch mark (see the preceding photos). When that was done, you aligned both with the sun as if a real eclipse was happening. The scale effect of the bead representing the moon would cast an umbra/penumbra combination that perfectly matched the profile of an actual solar eclipse on Earth itself.

A CROWD INTENSIFIES

After explaining the aspects of *Megamovie*, as well as demonstrating the Yardstick Eclipse to the few that were on our ridgetop, we quickly noticed that our ridgetop vista was beginning to experience a sudden, yet temporary, population boost. With each new visitor, the process of explaining the project (and being asked "Who's the Google Guy?") would begin anew.

By about two o'clock that afternoon, quite a few other eclipse watchers had arrived. One of the groups was from Seattle, and another had driven all the way up from Berkeley, California. Another one had driven over 2,000 miles, coming in from Boston, Massachusetts and one had come from as far as Amsterdam. By now, our rig (and my imposing setup of telescopes) was quite the talk of the ridge.

At around that time, our friend Jami from California, had called. The night before, he had got lost trying to find us. I think I had spent about an hour trying to talk him up our ridge, and after a few failed attempts he ended up spending the night on a nearby ridge. When he called, I was soon surprised to learn that his car was visible in a direct line of sight to our south on a neighboring ridge, so I "waved him over" using the same folding TV tray table I had painted white weeks prior. After a half hour had lapsed, he had arrived to join us on our ridgetop viewpoint. A fellow imager, he had the goal of photographing the eclipse as well as I did, and had set up two of his own telescopes to accomplish that. It wasn't long before it started looking like an impromptu "star party" (a gathering of astronomers with their telescopes) was taking shape.

As our friend Jami finished setting up his telescopes, a friend of mine, whom I went to high school with had also arrived on our ridgetop viewpoint. Within an hour or so, around 15 cars populated the top of that lone ridge. Before long, we were basking in views of the sun through a whole assortment of telescopes. Jami had two of his set up, plus I had brought down a Meade Newtonian for eyepiece views of the sun during the eclipse.

Late that afternoon, we got a call from our friend in Snohomish, who had camped overnight in a field in Goldendale, Washington (the same one we had initially planned on staying at). He was on his way, with yet another telescope (his plans were, like mine and Jami's, to photograph the eclipse). He was calling to let us know that he was hitting traffic north of town, and that it would be a while before he would arrive.

As the day progressed, everyone kept marveling at all of the telescopes, and those two huge *Megamovie* banners. It seemed with each new arrival (before dusk, an additional 10 showed up), the first order of business was a demonstration of the Yardstick Eclipse tool and an hour long demonstration of solar eclipse science.

Our ridgetop vista begins to populate with more eclipse watchers in the late afternoon hours of August 20, 2017.

AN ECLIPSE TO REMEMBER

Our friend Jami from California, has arrived and begins setting up.

A late evening sky, taken at 7 p.m. on August 20, 2017.

Sunset quickly approached, and the sky to the east darkened with approaching twilight. Simultaneously, the sky to our west turned a brilliant shade of vermillion red. It was the most spectacular sunset we experienced on the trip down. All of us remarked on that ridge, how long this sunset's colors seemed to last.

Not long afterwards, our friend from Snohomish arrived, and began setting up his scopes in anticipation of the eclipse. By now, the eclipse was only twelve hours away.

As night fell upon our ridgetop vista, we held an impromptu star party. Our friend Jami and I set up our scopes on various deep sky objects, while our friend from Snohomish set up his scope for a session of deep sky astrophotography on the Andromeda galaxy, which was almost directly overhead. Together, we shared views of galaxies, nebulae, and distant star clusters. The experience of that was exceptionally memorable.

As I sighted in a deep sky object known as the Lagoon Nebula (a rich summer target with a lot of nebulosity in the constellation Sagittarius), I alerted a few nearby to have a look.

One observer, a young child around nine years old, was awestruck.

"What's that bright cloud next to those stars?" he asked.

I replied "That is what they call a nebula, and this one has a name called the Lagoon Nebula because it kind of looks like a lagoon in photos."

"What's a nebula?" the youngster responded.

To which I answered, "A nebula is a cloud of gas and dust in space, that is lit by nearby stars or emission energy. Sometimes, they're also caused by dying stars."

After the Lagoon Nebula, I sighted in a few more targets, including the Andromeda Galaxy. Upon sighting in the Andromeda galaxy, one individual remarked "I can clearly see the spirals! This is incredible!"

Before long, everyone was afforded a view through all of the telescopes on that ridge. The experience for many was unforgettable, especially for the children. While our telescopes were aimed at the Dumbbell Nebula, we began explaining the dynamics of what a planetary nebula is. While doing so, one of the ladies in the group exclaimed "This is the most incredible night of my life! I never imagined having a view through a telescope like this! I will cherish this night for the rest of my life!"

By now, it was almost 2 a.m. Sunrise was fast approaching with just four and a half hours to go. We quickly shut our telescopes down, then charged our remote battery packs for the eclipse.

Before we turned in for the night, we remarked at how intense the fires were burning, over at the Milli Fire by the Three Sisters. This night, was the most intense they had burned during our stay, and the glow was bright enough to wash out on long exposure photographs. One thing was certain: If clouds weren't going to steal our show, the risk of being clouded out by wildfire smoke was becoming very real.

In fact, we noted that some of the smoke off in the distance was being illuminated by the lights of the city of Bend, Oregon, situated about 50 miles to our south.

TOP: On the eve of the eclipse, a vermillion sunset blazes over Madras.

BOTTOM: The Milli Fire near the Three Sisters volcanic chain rages on.

6

TOTALLY ECLIPSED

Eclipse day dawns on Mount Hood. A little more than four miles away, the center line of the path of totality crosses the field of view in this photo, at the bottom of the frame.

Taken at sunrise on eclipse day, smoke nearly decapitates our view of Mount Jefferson.

It would be a stretch to say we slept well the night before the eclipse. In fact, to be honest we hardly slept at all. Part of that was excitement, and for me, part of it was nervousness and anxiety.

Almost on instinct it seemed as though nearly every one of us on that ridge had woken up at sunrise, which on that morning was a little bit past 6:20. Our spirits were surging and excitement was quite high.

In this scene an hour after sunrise, Christine Rosenow gives a thumbs-up as eclipse-day preparations begin.

Our friend Jami woke up soon after us, and set up his portable coffee pot and camp equipment. Within 30 minutes we were all sipping back on a nice cup of hot coffee, on that icy cold ridge. Temperatures were down into the 40s that night, and when we woke up, the air still had a chill to it. It didn't take long, however, for that chill to fade into warmth as the sun slowly rose higher in the sky.

To our west, we noticed that the smoke column from the fires near Three Sisters had flattened out in a large "pancake" and a shelf-like projection of the smoky haze had drifted north and gave the appearance of having sliced the summit off of Mount Jefferson. As the sun rose higher, our attention was towards that smoke, and for good reason: It was drifting right towards us.

By 7:30, most of us on that ridge had finished eating our breakfast. The first morning cup of coffee had turned into our third, and fourth. Over at our telescopes, things were working smoothly. We switched the telescope tracking motors on at 8, followed by the cameras at 8:30, for a brief last-minute testing run. The control laptop was fired up and online by 8:40.

A half hour before the eclipse was predicted to begin, everything seemed as though it was going smoothly. Then the unthinkable happened.

AUTOMATION FAILURE

Up to now, everything was working smoothly, save for the automated scripts which we would be using to control both Nikon DSLRs. The eclipse was predicted to start its first partial phase at a little past 9:06 a.m. Fifteen minutes before that, the automated scripts in DigiCam Control did something I was hoping would never happen: *THEY STOPPED WORKING*.

Both cameras would now have to be controlled via a manually-controlled override feature in DigiCam Control, meaning that we would have to watch things on the screen full-time. It also meant we'd have to sacrifice observation of the eclipse through our Newtonian for the sake of making sure everything worked. Once the manual override was activated, the cameras would then begin automatically capturing partial phase images every five seconds, starting ten seconds before the clock struck 9:06.

By 9 o'clock, all of us on that ridge heaved a collective sigh of anguish. The thick shelf of smoky haze that had almost sliced Mount Jefferson in half at sunrise, was now directly overhead and drifting fast toward the sun. With five minutes to go until show time, there was no time to change exposure settings.

FIRST CONTACT

A last-minute decision was made to continue as planned, and the cameras started collecting images on time, and as planned, at 9:05:50. For the first several dozen captures, the smoke haze was obvious in the images coming off the camera. The sun, which under normal exposure settings in a white-light filtered telescope can appear bluish-white or orange (depending on the filter used), had now looked deep red. A feeling of deep concern washed over, thinking that this might be a bust. That feeling was even more pronounced when I realized that exposure settings could not be changed or adjusted mid-capture.

AN ECLIPSE TO REMEMBER

A few minutes later, we began to notice the first visible hint of change as the sun's disc was slowly being eclipsed. As we were noticing the first moments of the eclipse, the first few images were coming in off the cameras (there was a 20-second delay between capture and upload). Ten minutes into the eclipse, that delay ran well in excess of a minute. By now, it's about 9:15 and at this point, the smoke haze was starting to clear out. As the smoke haze cleared, the images coming off the camera were now looking more normal with each new capture. Fifteen minutes later at around 9:35, the sky was beginning to take on a more slate-gray tone to it. Visually, with quick naked-eye glimpses of the sun, you could tell something was biting at the solar disc. With solar eclipse glasses however, it was becoming more obvious that totality was fast approaching.

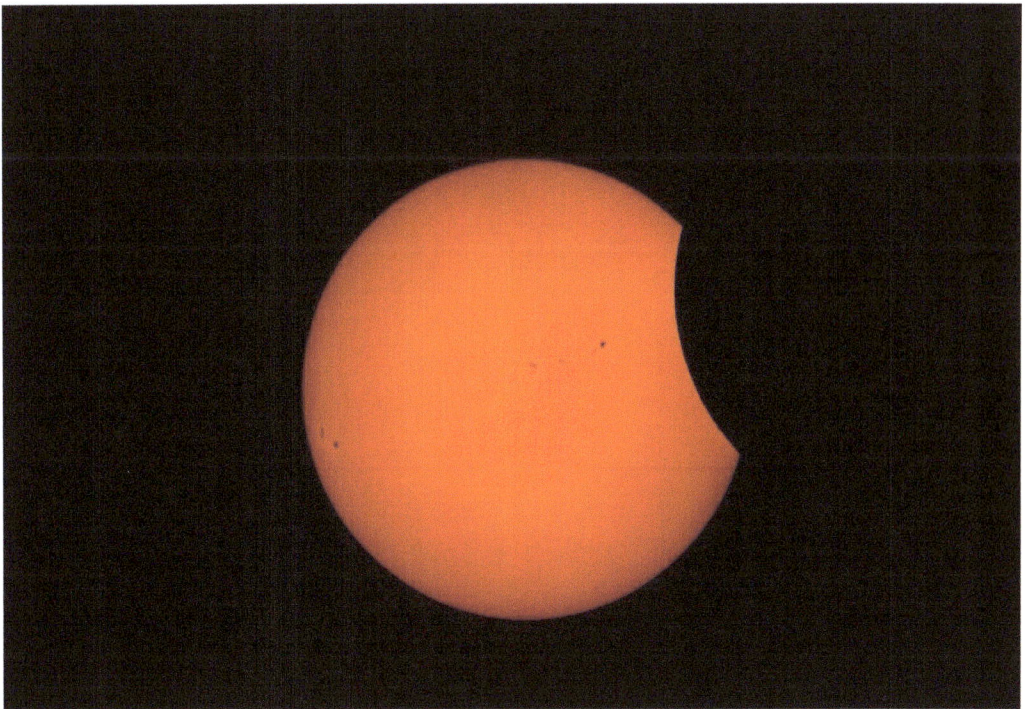

The first partial phase, taken ten minutes after the eclipse began at First Contact.

For the next half hour, the light began to gradually diminish, as did the color in the sky and the landscape. The changes were small, minor, and almost imperceptible at first, then began to change more rapidly as time progressed.

Nineteen minutes before totality – at 10:00 a.m. – the first real, ominous signs of change were beginning to unfold. To the west, the skies were darkening more quickly, and what had been shades of azure blue sky had been replaced by darkened shades of muted blues, purples and grays. It was especially noticeable in the skies above Mount Jefferson. To our west however, the skies still looked relatively normal in brightness, although the loss in color saturation was especially noticeable.

At 9:55AM, the sun began to take on a *PacMan*-like appearance. By now, the skies were beginning to darken considerably.

The real show, began within ten minutes of totality. At 10:09 a.m. we began to remark at how dim the sun was now shining, and once-blurry shadows were now becoming razor-sharp. Almost immediately, it seemed like the light of the sun was becoming similar to that of an LED-powered flashlight.

AN ECLIPSE TO REMEMBER

Then, the darkening of the sky began to happen rapidly, with almost all color saturation being simultaneously turned off. It was as if we were standing on the set of *The Wizard of Oz* and someone reached over and turned the TV's brightness knob down, and the color knob to black-and-white. The rapid loss of sunlight in the last few minutes seemingly happened in an instant, and unfolding before us was the greatest spectacle of all.

At around that time, I suggested that if anyone had a white sheet with them, to grab it just in case "Shadow Bands" might appear (a feature of shimmering light often seen during eclipses right before and immediately after totality). One of the guys ran to his car, and came back up with one just in time. As it turned out however, we were saddened to find that shadow bands would not be seen from our location.

One minute away from totality, the sun had been reduced to a sliver.

Twenty seconds before totality, one of the individuals in a group with us on that ridge looked to the west and noticed Mount Jefferson had all but disappeared. "Where's the mountain?!" exclaimed another. We immediately looked west, and noticed the profile of Mount Jefferson stuck out as an inky black silhouette against a dark sapphire blue/gray sky.

At that point, I exclaimed "Mount Jefferson is dark! 17 seconds to totality!" as a warning that totality was imminent. As we were watching, the valley before us began darkening at what seemed like a breakneck pace. The sky above us was darkening rapidly as well, at almost lightning speed. The remaining sunlight too, was rapidly disappearing. Within seconds, Baily's Beads began appearing, and the light of the sun would be snuffed out as though it were the flame of a flickering candle. The last bead shone brilliantly in what astronomer's call "The Diamond Ring effect." In a split second, the skies suddenly went dark.

SECOND CONTACT

The total phase of the eclipse started 42 seconds past 10:19 a.m., as the skies were darkening above us. The skies in those first few moments were almost full-moon dark. Our cameras, which had to that point recorded a combined total of more than 1,600 frames, had been switched to full manual override. Above us, was the brilliant spectacle of a total eclipse: The first diamond ring. Within seconds, the diamond ring had faded and totality was upon us. For the next two minutes and five seconds, the greatest spectacle of a totally eclipsed sun was playing out above us.

The solar corona – the twisted, fine tendrils of the outer solar atmosphere with temperatures of more than one million degrees – was shining brilliantly like strands of angel's hair lit from behind. The inner corona's brightness was especially remarkable, shining as bright as a full moon. Three distinct coronal lobes were visible. Where the sun's disc brilliantly shone an hour before, had been replaced by a deep, dark blue 'dot.' Earthshine, the light of a fully-sunlit Earth bouncing back off the moon (sometimes seen during thin crescent moon phases), was making lunar features faintly visible to the naked eye. To the left, the star Regulus, the brightest star in the constellation Leo, was shining brilliantly, as was Nu Leonis almost equally distant to the right. Directly overhead, the planet Venus shone brilliantly like a diamond in a sea of indigo blue. To the south, the star Sirius in the constellation Canis Major was piercing through the twilight.

The sky above our site at mid-totality. The moon's umbra darkened the smoke haze to hues of dark brown. The bright "star" at top right center is Venus.

Photo captured from GoPro video courtesy of Jami Boothe.

All around us towards the horizon, the sky took on a most unusual hue of grayish-orange, like the light of an early morning before sunrise. To the north of us, and just outside the path of totality, Mount Hood and the far northern reaches of the valley to our north, shone with the color of faded alpenglow, as did the Three Sisters and distant ridges to our south. All the while, we were basking in the darkness of totality. To our west, we were stunned to witness the speed of the moon's umbra, darkening distant smoke haze as though it were like a curtain coming down on a brightly-lit stage.

Standing at our location, you could clearly see and detect the speed in which the umbra was traveling. At the height of totality, the sky above us took on a deep shade of blue, with hints of silver and purple. It was the most stunning display of color we've ever witnessed. The brightness of the darkest part of the sky, was no brighter than the light of a fully moonlit sky. Along with the appearance of a moonlit sky, the temperature change was also noticeable. At the height of totality, the temperature had dropped nearly 21 degrees.

By now, each camera had captured more than 30 combined exposures across several bracketed exposure levels. Each set of bracketed exposures were fired off by depressing both shutter buttons on the cameras manually. In between sets of three, the exposure levels were changed one stop by rotating the camera dial for the next exposure set. As we were capturing the eclipse and the dynamic range of the solar corona, my wife and I were looking up in awe. This show was definitely not one to miss.

About halfway through the total phase of the eclipse, the first images of totality started transferring in from the cameras. Since there was a lengthy delay at this time, the images of the first few seconds weren't filtering through until almost halfway through totality. A group of six observers on the ridge stood behind us at our computer as we were watching them come in. As those first few images started displaying, a few of them screamed in excitement. Emotions were running high, and feelings of pure joy and elation were had by all who witnessed it.

Especially noticeable was a bright reddish finger-like extension of the solar chromosphere - called a 'prominence' – had stuck out from behind the moon like a giant sore thumb. It had the intensity and appearance of a bright pinkish-red neon light. Soon, that prominence was joined by a fainter arc of chromosphere peeking from out behind the moon. It was as if there were a giant red neon arc at the edge of the moon. The appearance of the chromosphere meant totality's end was soon near.

Twenty seconds before totality ended for us, it was coming to an end on Mount Jefferson. Its previously inky black silhouette was now basking in slowly returning sunlight. Everything that led to totality was playing out in reverse before our eyes. Within seconds, the clock would strike 10:21 a.m., and the second Diamond Ring would signal the end of totality in what many were calling "The Great American Eclipse."

AN ECLIPSE TO REMEMBER

Totality, from two different perspectives. At top, stars become visible in the wide-angle photo. In the bottom perspective the finer tendrils of the solar corona take on the appearance of fine hairlike strands – each one tangled or frayed by solar magnetism.

THIRD CONTACT

At third contact, the moon's umbra began quickly moving off to our east. The boundary between the umbra and penumbra portion of the shadow was especially prominent as the Diamond Ring effect began to show.

Photo captured from GoPro video courtesy of Jami Boothe.

The piercing light of the photosphere peeking out from behind the lunar surface creates the Diamond Ring effect, seen here at Third Contact. In this photo, the solar corona still shines brightly against rapidly returning sunlight.

Christine Rosenow photo.

AN ECLIPSE TO REMEMBER

The eclipse's totality phase ended at precisely forty-five seconds past 10:21 a.m., announced by the second Diamond Ring, or what astronomers officially call "Third Contact." Everyone around us was elated, and tears of joy began to flow. Right after totality ended, we resumed the manual override control in *DigiCam Control* and our cameras began capturing images every five seconds once again. By now our cameras had captured more than seventy-five bracketed images of totality, and just under 1,700 photos in all.

One thing was certain: All of us on that ridge, had witnessed the best show on Earth.

The first order of business following totality was clear: 1.) Upload the initial raw data files to the *Eclipse Megamovie* data servers, and 2.) Import that data to Adobe Lightroom for initial post-processing. Since the eclipse was being captured in Nikon's NEF RAW format (for the highest quality data possible), it was imperative to post-process all that data. That way, it could be exported before it could be posted on the web. After processing the first few photos of totality, a WiFi hotspot was activated so that uploading to the internet could commence.

The eclipse's final act has begun. This photo was taken roughly twenty minutes after totality ended.

Two minutes after totality ended, our images of the eclipse were going live over the internet. A few seconds after uploading the first image to Twitter, the ABC affiliate in Seattle, (KOMO-TV 4) immediately picked it up and retweeted it. Within ten minutes, our first photo of the eclipse was being used as the lead image in a KOMO story, which was authored by producer/meteorologist Scott Sistek. As that was happening, we received a confirmation e-mail from the *Megamovie* team, saying our data were received. As the *Megamovie* team planned on doing a live stream with a preliminary draft two hours after the eclipse would end in South Carolina, (or roughly four hours later) uploading that data was especially important.

With a half hour remaining, the eclipse's final act draws to a close.

Several minutes after the totality phase ended, we noticed the sky began to brighten faster than it had darkened in the moments leading to totality. As this was occurring, we also noticed that the temperature began to rise quite quickly. For the next hour and 20 minutes, most of us on that ridgetop viewpoint could not stop talking about what we had just witnessed. Others were taken aback, and were solemnly reflecting on what they had just saw. My wife and I shed tears of joy about 20 minutes after the eclipse ended. This was our first eclipse. It was at that point, we had mutually decided that April 8, 2024 was going to be our next one.

AN ECLIPSE TO REMEMBER

The eclipse was predicted to complete its second partial phase nearly an hour and a half later, at 11:41 a.m. Immediately after totality ended, a few on the ridge began packing up, and within twenty minutes after the end of totality about one-third of those on the ridge had already left. Our goal was to capture the eclipse from start to finish, so we elected to stay. By that time, over 2,500 images had been recorded so far, with about another thousand to go.

About twenty-five minutes after totality, temperatures were also climbing into the 80s again, and most of the sunlight had returned to what seemed normal to us. The sky to our east no longer looked dark, although we did notice a thickening in the haze from the nearby Milli Fire.

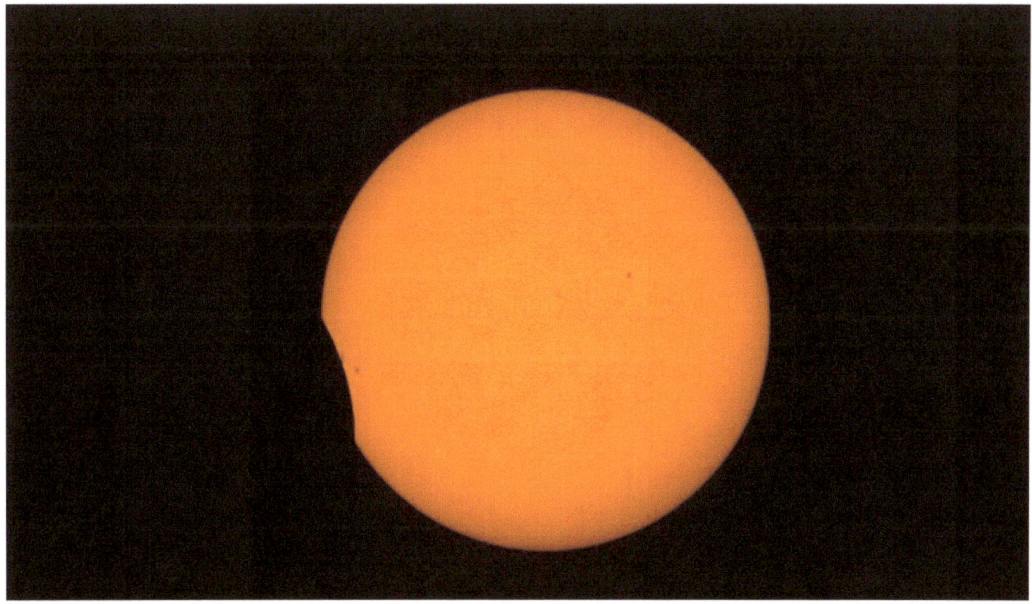

With just two minutes remaining until the end of the eclipse, the lunar disc begins to ease out of the frame. By this time, the sun looked normal to the naked eye.

As the eclipse was winding down and brightness had mostly returned, the stark reality of nightmarish post-eclipse traffic was unfolding before our eyes. In the valley below us, Highway 97 off in the distance was already at a complete standstill. To the naked eye, it looked like a string of tiny multicolored beads packed tightly close to one another. In binoculars, it stretched out across the valley as far as the highway was visible. Down below us, the farm that been previously a makeshift campground, had by then half emptied. Traffic on E. Loucks Road had taken on the appearance of being like Interstate 5 in Seattle at rush hour.

FOURTH CONTACT

The eclipse performed its final act on schedule at thirteen seconds past 11:41 a.m. Two minutes after the eclipse had ended, I stopped the capture routine on the cameras and switched DigicamControl down. That was soon followed by powering down the telescope mount, and both cameras. Temperatures were now crawling into the upper 80s, and that made for packing our gear somewhat unbearable.

After soaking in the entire experience, we then began the long process of breaking our campsite down. It would take an hour, deflating our airbed, followed by folding our tent back up into its carry case. Before the tent and airbed could be packed into our SUV, we carefully had to repack 150 pounds of telescope and camera gear into the back end of a 1996 Ford Explorer. Packing the telescopes, our tent, air bed and supplies took about forty-five minutes, followed by a few final farewells to those who had stayed up there with us.

After the eclipse ended, a feeling of sadness had set in, likely because the following day we'd all have to return back to a routine of normalcy. After having everything packed, we were now ready for our journey back home. First, though, we needed to get something to eat. This time, it wasn't going to be sandwiches and granola bars.

PREPARING TO LEAVE

As we were leaving, we noticed that charging our mobile devices using our car's battery during the entire time we were on that ridge, had the unexpected side effect of having a dead battery. Since we couldn't start our vehicle, we initially thought that we'd be stuck in Madras with no way home. No one that was up on that ridge with us, had a set of jumper cables.

After failing to start our SUV, someone in our group recommended that we ask the people down in the customized RV below us. After walking down to where they were, they graciously loaned us a pair of commercial-duty jumper cables. Since the kind they provided were commonly used to provide charging power to lifting equipment (such as electric forklifts) and serve as tractor-trailer jumper cables, they did the trick instantly. After starting our rig, we then waited ten minutes for the battery to recharge.

While we began to leave, I noticed that the traffic jam on E. Loucks Road had cleared somewhat. The makeshift campsite on the farm to our north had largely emptied out by that point, and most of the travelers heading out from that had largely made their way to Highway 97 or other outbound arteries. For us, we were still an hour behind. Our friend Jami agreed to meet up with us at a Taco Bell in downtown Madras. From there, we decided it was best to leave. Interestingly, when we collectively left as a group, the only people remaining on the ridge were the people that were initially there when we arrived two days prior.

FAREWELL, MADRAS

Our first stop, after descending down the ridge, was a wall of traffic on East Loucks Road about five blocks east of Highway 97. From there, it was pure gridlock. Taco Bell was a little under a mile away, and it seemed like it would take us forever to get there. When we got onto Highway 97 from E. Loucks Road, traffic was at a snail's pace until we got to the junction with Highway 26, where we would cross the intersection and continue south for a few more blocks. With traffic moving no faster than walking speed throughout the downtown core, it took us a little more than an hour just to get there.

When we arrived, placed our order, and then sat down. The expected wait time was about 20 minutes for our order to arrive (a delay which was due in large part because of everyone else doing what we were doing – waiting it out). While we were waiting, we noticed an unusually high police presence. At first, we thought a traffic stop had occurred that led to a foot chase. Drivers began exiting their vehicles and one of them began talking to a Jefferson County Sheriff deputy and a Madras police officer. A few moments later, a customer walking in the door adjacent to us, informed us that the bank next

AN ECLIPSE TO REMEMBER

door to the Taco Bell had been robbed. We both exclaimed "Hey that explains the two dozen cops!"

By this time, cops had surrounded both buildings, and were blocking the only way out of the Taco Bell parking lot.

After ordering, we waited for what seemed like an eternity for our friend Jami to arrive. As we were eating, we took the advantage of using the free WiFi to upload more photos of the eclipse to social media (as well as uploading more data to the *Eclipse Megamovie* servers). After 45 minutes, he arrived. At that moment, we exchanged views of our data and reflected on our collective experience as a group. We couldn't stop talking about it. Not long after, we all agreed that it was time to leave. Our friend Jami left first, followed soon after by my wife and I. We decided the best course of action was to follow his lead and make a run for it. It was not going to be easy.

Seen from an overlook, travelers heading south on Highway 97 were mired in endless gridlock.

In fact, getting out of Madras initially looked like it might be impossible (at least on that day). Traffic on Highway 26 and Highway 97 – both of which constitute the main north and southbound routes in most of downtown Madras – was at a standstill. The gridlock was so severe, that Madras – a town of 6,400 – may as well have been a city like Seattle at rush hour. Any law enforcement that weren't involved in the bank robbing investigation, were stationed at strategic intersections to direct traffic, as were a large number of local firefighters. Even the National Guard had a strong presence guiding traffic in the area. The additional "guidance" and efforts to improve traffic flow turned out to be an effort in futility. When we departed Taco Bell, our next stop was a Safeway gas station a half mile away. It took us more than an hour to cover that distance. After getting gas, our next stop – hopefully – was home. By normal standards, "home" was a little more than four and a half hours away. However, that was not to be the case.

We left the Safeway gas station at 1:45 that afternoon according to the in-vehicle clock in our SUV, and our cell phones. Immediately after exiting the Safeway parking lot, we were met with stopped traffic for miles. The first wave of gridlock we hit, took us a little more than three hours to cover the 1.25 miles from the gas station to the Madras Airport entrance and the entrance to the SolarTown encampment. There were still literally tens of thousands of people

Beginning our ascent out of Madras. By the time this photo was taken, we had already been mired in traffic for more than two hours. Up ahead, the Madras Airport was less than a mile away.

still camped out there, with hundreds more trying to leave and occupy the same stretch of highway we were on. As we were passing SolarTown, we witnessed several television crews doing live reports. The same road we experienced gridlock on that prior Friday afternoon, was backed up for more than two miles alone in both directions.

Exhausted from gridlock, drivers and passengers alike take a break. Seen here in these two photos, a woman performs stretches, while up ahead others take advantage of the stopped traffic to grab refreshments.

From the entrance to the Madras Airport, it would take us another four hours to get from there, to Warm Springs. The distance was just sixteen miles. In most circumstances, that distance could be covered in about fifteen or so minutes. The speed in which we were traveling for the most part, never exceeded five miles per hour.

Most of the time, we would be sitting at a standstill. As we were approaching our descent into Warm Springs, we saw a television crew on the side of the road doing a live report. At first, it looked like a crew from an Oregon television station. As we passed, we would be surprised to learn that it was actually a crew from our local NBC affiliate, KING 5 Television out of Seattle, Washington. While driving by (and because KING5 was local news to us), we waved at them. As we took a snapshot of the scene, the reporter waved back at us.

Within a few miles, we began to descend down into Warm Springs. Upon our descent, traffic actually seemed to improve somewhat, and we quickened our pace. Up to that point, the fastest we'd been able to drive was just 5-10 MPH. We were celebrating the 15 MPH we were making. Before long, we were having a party when we finally broke 20 MPH. Traffic soon slowed again once we reached the town of Warm Springs itself, where we slowed once again to below 10 MPH. The slowdown was most likely due to a wildfire which was being extinguished on the side of the valley just behind a mill. As we were passing through, the air had a very strong, almost overpowering smell of smoke.

AN ECLIPSE TO REMEMBER

A few miles north of Warm Springs, traffic seemed to even out, and for the next nineteen miles we were surprisingly able to travel at normal highway speeds (the posted limit of 55 MPH). At first, we thought we were in the clear. By then it was almost 7 p.m., and we figured that we'd be arriving at home a little before 11 that night. I figured that we'd be rolling through the Mount Hood area around sunset, and had planned to take a series of sunset photos on the Timberline Lodge-side of the volcano.

Our departing view of Mount Jefferson was mired in thick smoke and haze from the Milli Fire, which had thickened throughout the day. Like our arrival, the air had begun to smell of smoke as we approached Mount Hood.

STALLED AGAIN

About seven miles before the junction of US Highway 26 and Oregon State Highway 216, traffic once again slowed to a crawl. We would be hitting our second wave of gridlock at around 7:45, and this time, it would take two and a half hours to cover a distance of twenty miles. The wave of traffic we hit here, we concluded, was the first wave of people to leave Madras. They all left as totality ended, unlike many (including us) who stayed until the final act at

11:41 that morning.

As we were stuck in the second phase of gridlock, it became quite an amusing sight to see campers and cars, and vehicles of all kinds pulled over to the right shoulder, their owners having brought out camp stoves and cooking equipment. Slow vehicle pull-offs became impromptu campsites. With the gridlock having no end in sight, it seemed as though the shoulder of the highway became the ideal location for that night's dinner table.

Traffic once again reaches a standstill in this photograph taken approximately 20 miles from Mount Hood.

Even more amusing, was the action on our left. Not far from the Clear Lake area, we decided to switch to the right lane out of courtesy to a group of people behind us. After we did that, traffic came to a complete stop, and a group of people in the car to our immediate left initiated a game of 'Rock-Paper-Scissors' that lasted for an hour. Another thing that was unique, was the fact that people actually held conversations with the people in the cars next to them. For us, many asked what kind of photos we took. Often, conversations could last more than 10 minutes before traffic would creep ahead, then it was onto the next car.

After making our way closer to Mount Hood, traffic had slowly begun to improve. Traffic was now improving to the point that when we reached Government Camp and Timberline at Mount Hood, we were traveling at roughly half the posted speed limit once again. It would be about thirty-five more miles before we were able to consistently follow the posted speed limit. By about 11:25 we were heading towards Gresham once again. According to our GPS, arriving at home was still two and a half hours away.

Descending into Portland, we first took a slight detour before Gresham and passed through the town of Boring, Oregon. Our goal, since we had traveled for quite a length of time by that point, was to find a casual-dining restaurant to have a real meal. Unfortunately for us, we didn't find many that were open (and those that were open, were packed), so we continued on and decided it was best if we tried to find one across the border in Washington.

Upon crossing the Columbia River on Interstate 5, our first stop was in Vancouver where we tried our previous approach of finding a decent casual-dining restaurant. We were disappointed to find that the restaurant we stopped at, was closed despite an overwhelming presence of customers in the main lobby and in its parking lot. We didn't really feel up to fast food at the time, and had called the restaurant to see if they would be willing to make an exception for us. We used the plea that we'd been driving for more than twelve hours by that point. However, the plea effort would fail, as the manager on duty was unwilling to open up the doors.

After we left, we decided it was time to give in, and realize that fast food was our only viable option at that point. A few blocks away was a McDonalds, at which point we stopped and ordered off their menu. Since the dining room and lobby was closed, we ended up eating in the parking lot of a nearby Safeway. After eating, we disposed of all of our garbage in a nearby trash receptacle, and immediately left.

Upon leaving and making our way to our final leg home, our first order of business was stopping at the Gee Creek Rest Area on Interstate 5. The first and foremost reason was that by that time, we badly needed a restroom break after all that driving. The second reason was that we needed to empty our portable toilet into their RV dump station.

One thing we quickly learned, is that deodorizing chemicals or 'scent blocks' don't work for very long, and hot weather can speed up the process in which one of those portable toilets can begin to stink. By the time we got there, the odor of the bucket's contents (despite being completely sealed) was becoming quite nauseating. After emptying it, we hosed it out for good measure.

After leaving the rest area, we proceeded north yet again. Not long into the last leg of our drive, we hit another wave of eclipse traffic on northbound Interstate 5. This wave of traffic, not unlike the first two we encountered, took us quite a while to cover a short distance. The jam began just south of Napavine, and it took us just over an hour and twenty minutes to cover a distance of just twenty miles. We cleared the congestion just north of Grand Mound, Washington. As soon as we cleared the gridlock, our GPS alerted us that home was just an hour away.

We would soon discover that the gridlock we cleared, was the last wave of eclipse traffic that left the west side of Oregon. Upon passing Grand Mound, the rest of the drive was without incident. Our last two stops were at a 7-Eleven in Tumwater for a small amount of reserve fuel, and a stop off at Walmart back in our hometown of Shelton. After Walmart, home was just a mere 10 minutes away. We then shut our GPS off, and made the final leg of our journey. At a little past 2:30 a.m., we pulled into the driveway, home at last.

Exhausted, tired, and with sore backsides from 13 and a half hours of driving, we were excited to be back home. It was a fantastic, fun, and event filled conclusion to an awesome and epic experience.

AFTER THE ECLIPSE

Our experience at the eclipse down in Madras, was simply unforgettable. In the first week after the eclipse, the first task was polishing up the high-dynamic range composites of the solar corona. The process of creating the final two images involved over 200 hours of processing more than 70 separate photos. Since two telescopes were used at different focal lengths, it was especially tedious.

Each set of images was subsequently divided and stacked by hand in Adobe Photoshop to create a master layered file. More than 30 layers, and five groupings, were used. Each layer (which was its own image) was then deconvoluted, and a special processing routine known as the Larson-Sekanina method was applied. Larson-Sekanina processing methods are more commonly used in photos of comets to bring out details within the coma and cometary tail, however in this instance they would bring out the finer details of the solar corona (details like coronal streamers and magnetic loops of coronal plasma).

Afterwards, each layer was blended into a duplicate of itself. Finally, a variation of the same processing method used by Miloslav Druckmüller was then applied. Upon the completion of all that editing, the photos were then readied for the internet, and ultimately for commercial sale through our website.

Once the totality images were processed, stood the monumental task of creating the two planned time lapse videos. Each time lapse video contained over 1,600 combined frames for a seamless video of the event. As we were capturing the eclipse we ensured that during the partial eclipse phases, the settings would not exceed five captures per second. This was partly because our cameras were high-resolution cameras (which would ultimately write a ton of data, and consume a profound amount of hard disk space), but partly because with the time lapses we had planned, five second intervals were all that were required.

The process of compiling each time lapse began with the daunting task of separating over 3,100 frames in Adobe Lightroom, by way of separating each eclipse phase by a star-based rating system. Once that was accomplished, the photos were batch-exported and then subsequently loaded into Adobe After Effects, where they would be motion-stabilized (to account for movement of our telescope mount during the eclipse).

Following that, separate images (including the Baily's Beads sequences and the sequences of totality) were compiled into their own separate videos using the same process for the partial phase videos. Finally, each set of videos was compiled as a single video of approximately one and a half minutes in length. Both videos would later be incorporated into a single video 'feature' which showcased both videos as one complete compilation.

Not long after all of our data was finalized (and our time lapses released to the internet and social media), came something unexpected: Doing public speeches about our experience. Our videos and the two photos of totality were a hit on social media, so it seemed natural that we'd be asked to speak about it.

AN ECLIPSE TO REMEMBER

The first public speaking engagement came about in an interesting fashion. It's also one which I am extremely proud of.

Three weeks had lapsed since the eclipse, when our niece Julie had called. As it turned out, her freshman science class was in the process of learning about the solar system and the dynamics of stellar formation. After wrangling with planning and having to sign clearance forms for the Bremerton School District, I was formally invited to speak.

On September 25, we traveled to Bremerton High School and spoke to a group of around 30 freshmen. Initially we thought our speech would be met rather coldly (after all, some of today's youth seem rather uninterested in science). However, the entire class gasped in awe at our time lapse video, as well as some of the software I was displaying. As we later discovered, speaking in front of a group of high school students was something that came naturally.

A few months after the eclipse, a friend of ours (who had initially suggested we try Madras instead of Mill City), had asked us to speak at a monthly meeting of astronomers with the Everett Astronomical Society. The privilege of being asked to speak among a group of fellow astronomers is one of the highest honors one could ever experience.

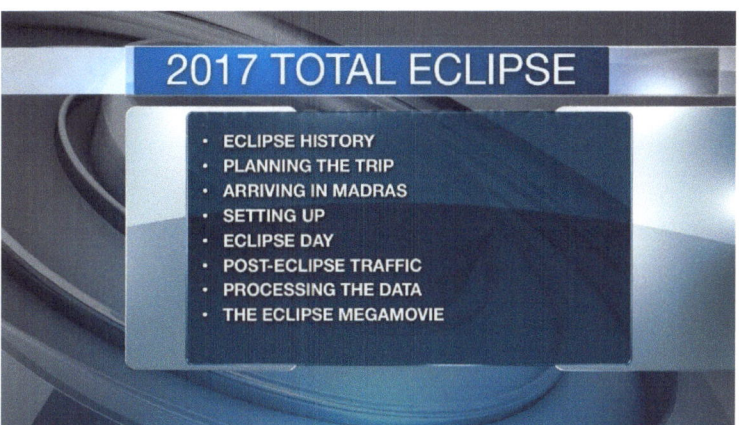

A slide from one of our PowerPoint presentations on the eclipse. This was used for our speech at the Everett Astronomical Society.

Our experience at the solar eclipse also opened up a bit of a windfall. Until the eclipse, most of our photographs (and those in which we've sold as wall art) were mostly of the fine-art scenic and landscape kind, interspersed with a few wedding and family portrait shoots. Additionally, we've sold a number of our astrophotography-related photos. When we first released our final eclipse composite, we were surprised by how many purchase orders came in.

Additionally, we would be surprised to learn that demand for eclipse photographs was quite high. Due to the popularity of our photo on the photo sharing site *Flickr*, the editor and publisher of an e-zine titled *Amateur Astrophotography* had decided to use it on the cover of their September issue.

It would also become our highest-grossing photograph we'd ever sell, to-date. Available in sizes ranging from 8"x10" to as large as 24"x36" poster size, more than five dozen prints of that photo sold. Our most popular size ended up being a matte-framed 11"x14" version that came framed in stained mahogany.

Nothing however, can convey the experience of actually being at a total solar eclipse. Watching videos of a solar eclipse just cannot compare to the actual experience of being there. Personally seeing the eerie change in color, the sky darkening to the brightness of a moonlit-night in the middle of the day, and being able to see a large chunk of the solar atmosphere that we otherwise cannot see are experiences that are unforgettable and profound.

Seeing the eclipse (and since it was both our first), it turned both of us into what some call "umbraphiles." The term umbraphile is given to anyone who becomes an eclipse chaser. The experience of seeing an eclipse is one that is raw, full of emotion, and can be life changing. With many, it is not uncommon to become more philosophical about life after seeing one.

As we were driving home, we discussed the idea of traveling to see the April 8, 2024 eclipse. After reviewing the path of totality maps, we mutually decided upon trying to photograph and observe it at Enchanted Rock State Park, outside the city of Fredericksburg, Texas.

AN ECLIPSE TO REMEMBER

Ironically, the location we chose would be on the same relative moon-sun axis as our location in Madras. Even the lunar axis angle would be the same. That means that from our chosen vantage point on Earth, the moon will have the exact same lunar terrain profile as it did for us in the August 21, 2017 eclipse. With those factors, photographing the eclipse will likely result in an identical Baily's Beads sequence.

The August 21, 2017 eclipse carried so many memories with it. As it was our first eclipse, the memories we have of that trip are ones in which we'll cherish for a lifetime. Also worth cherishing were the friendships we established that hot August weekend. Since the eclipse, we've kept in contact with several of those on that barren ridgetop.

In fact, as the idea for this book was being fostered, we were watching a 1958 film about the Titanic disaster. The film's title, *A Night to Remember*, was based upon author Walter Lord's 1955 novel of the same name. At that point, watching *A Night to Remember*, plus every portion of our experience in Madras, Oregon brought forth a sudden inspiration for this book's title.

It was *An Eclipse to Remember*.

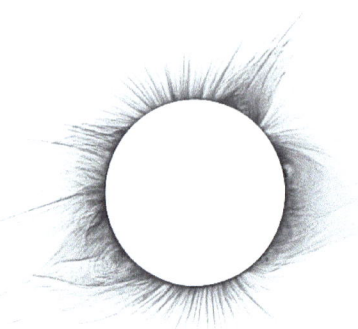

GLOSSARY

This book contains terms that need further explanation. As it was being compiled, I felt it necessary to explain those terms in greater detail.

ACHROMAT: A simple type of telescope which uses a convex and concave lens to form an image

APOCHROMAT (APO): A type of achromatic refractor which uses an additional lens element to refine the image.

BRIDGE CAMERA: An advanced type of point-and-shoot camera, with controls similar to, and having the appearance of, a DSLR (See; DSLR, SLR)

CATADIOPTRIC: A type of telescope which uses lenses and mirrors to produce an image.

CANON: A Compilation of eclipse predictions in table format, including times, dates, and coordinates.

CHROMOSPHERE: A layer of the solar atmosphere roughly 3,000-5,000 miles deep. Situated directly above the photosphere (the visible surface of the sun), it is only visible during a solar eclipse.

CORONA: The outer layer of the sun's atmosphere, which stretches outwards at a distance of several million miles. Its reflective light seen in total eclipses is caused by ionized solar plasma, heated to temperatures in excess of one million degrees Fahrenheit.

DSLR: A digital variant of an SLR camera (**See SLR**).

ECLIPSE: An astronomical event in which one celestial body obscures another.

MAXSUTOV-CASSEGRAIN (MCT): A type of catadioptric telescope which uses a spherical primary mirror and spherical corrector lens.

PROMINENCE: A projection of the chromosphere which extends outward beyond the chromosphere itself. Caused by magnetism.

REFRACTOR: A type of telescope which solely uses lenses to produce an image.

SLR: Short for Single-Lens-Reflex. A type of camera that uses mirrors and prisms to allow the photographer to see through the lens, thereby permitting the photographer to see what exactly will be captured.

SCHMIDT-CASSEGRAIN (SCT): A type of catadioptric telescope which uses a Cassegrain reflector's optical path, with a Schmidt corrector plate to make a compact astronomical instrument.

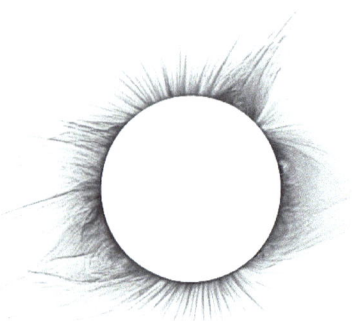

SOLAR ECLIPSE TERMINOLOGY

PARTIAL ECLIPSE: An eclipse in which the moon only partially obscures the sun. Total eclipses always begin and end as partial eclipses.

ANNULAR ECLIPSE: An eclipse in which the apparent diameter of the moon is less than that of the apparent diameter of the sun. In the central line of eclipse, the sun appears as a bright ring, or annulus.

HYBRID ECLIPSE: An eclipse in which the eclipse appears as both a total and an annular. Comparatively, this type of eclipse is rare. These types of eclipses begin and end as annular eclipses, while the central eclipse portion is a total eclipse.

TOTAL ECLIPSE: An eclipse in which the moon fully obscures the sun.

PATH OF TOTALITY: Anywhere on Earth that a total solar eclipse occurs, often more than a hundred miles in width and several thousand miles in length.

BAILY'S BEADS: The last few bits of sunlight peeking out through variations in lunar terrain during a total eclipse (or during portions of an annular eclipse). Named in honor of Frances Baily, who discovered them.

AN ECLIPSE TO REMEMBER

UMBRA: The innermost (or darkest) part of a shadow, where the light source is completely blocked by the occluding body. An observer standing in the umbral portion of a shadow would experience a total eclipse.

PENUMBRA: A region in which the light source is only partially obscured by the occluding body. In the penumbra, an observer would experience a partial eclipse.

FIRST CONTACT: The moment of eclipse in which the lunar limb touches (or contacts) the solar disc.

SECOND CONTACT: The moment of eclipse in which the totality phase begins. Immediately preceded by Baily's Beads and the Diamond Ring effect.

TOTALITY: The duration in which the moon completely obscures the sun.

THIRD CONTACT: The moment of eclipse in which the totality phase ends. Usually announced by a single point of intense sunlight (a second Diamond Ring), which rapidly spreads out into Baily's Beads.

FOURTH CONTACT: The moment in which the trailing edge of the lunar disc no longer contacts the solar disc.

DIAMOND RING EFFECT: The moment in which the last speck of Baily's Beads is seen, immediately before totality. Also seen as the first Baily's Bead at the end of totality.

In this photo taken five days before the eclipse, the authors are seen photographed next to the gear they would use (and attire they would wear) in Madras.

ABOUT THE AUTHORS

Steve Rosenow is an accomplished photographer, with published appearances in trade magazines and in print and broadcast. A number of his photos have been shown in documentaries, on local and national news, as well as a two-time appearance in *Ripley's Believe it or Not!* annual publications. An avid weather guru and astronomer, his contributions can be seen frequently on Seattle's KOMO-TV 4. He grew up in the Port Orchard and Purdy area of Washington state as a child, and spent his adolescence growing up in the small community of Shelton, Washington. From a very young age, he always saw the night sky as being full of wonder and amazement.

Christine Rosenow was born and raised in Olympia, Washington, having spent her entire childhood and adolescence in the same neighborhood. Like her husband, she shares an equal passion for nature and photography. She also enjoys cooking, and is known for making an exceptional batch of shrimp scampi. Her accomplishments include winning second-place titles in baking contests at the Thurston County Fair, and like her husband, has had a number of photos published through Seattle's KOMO-TV.

Christine and Steve married in January, 2015. Together they own a photography business known as **Loowit Imaging** and both are part of a group of photographers known as the Legion of ZOOM, which contributes content to Seattle's KOMO-TV and neighboring affiliates. They live in Shelton, Washington with their furkids Lucky, Lady, and Music, plus Christine's son Shawn.

www.ingramcontent.com/pod-product-compliance
Lightning Source LLC
Chambersburg PA
CBHW051152220526
45473CB00003B/750